Algorithms

Master key ideas in math, science, and computer science through problem solving.

Sign up for Free Now

Description

When faced with the word "problem," the reactions we experience are different. I remember in elementary school when the teacher used to write on the blackboard with catastrophe size letters PROBLEM, with its neat and rounded italics. He seemed to enjoy it. At that moment I could observe different effects among my colleagues, most of them accompanied by the gesture of taking the head with both hands. Others seemed to enjoy the challenge, among which today I include myself, although my memory may fail me a little.

But what was behind that statement? If we apply high levels of abstraction to the situation, we can surely affirm that what is proposed when enunciating a problem is that something be transformed, that certain things pass from state A to state B, different and desirable.

This book was designed for anyone who wants to start immersing themselves in the world of problems and algorithms, whether out of personal curiosity or within an academic program, using resources that I think are new and interesting that I have been developing through my years as a teacher. It is a foundation, in my opinion necessary to break with the myths that surround programming. Making programs is neither more important nor more complex than solving problems. That's the challenge, what changes are the characters and therefore the toolbox. The ability, will and talent are all ours.

Thanks for downloading this book. It's my firm belief that it will provide you with all the answers to your questions

Introduction

What are algorithms?

No doubt, you've heard the term before. It is often associated with all sorts of technical mechanics but in recent years algorithms are being used in the development of automatic learning, the field that is leading us to advancements in artificial and computational intelligence. This is a method of analyzing data in a way that makes it possible for machines to analyze and process data. With this type of data, computers can work out and perform a number of tasks it could not originally do. They can understand different concepts, make choices, and predict possibilities for the future.

To do this, the algorithms have to be flexible enough to adapt and make adjustments when new data is presented. They are therefore able to give the needed solution without having to create a specific code to solve a problem. Instead of programming a rigid code into the system, the relevant data becomes part of the algorithm which in turn, allows the machine to create its own reasoning based on the data provided.

How does this work?

This might sound a little confusing but we'll try to break this down into certain examples you can relate to. One of the 'learning' functions of machines is the ability to classify information. To do this, the input data can be a mix of all types of information. The algorithm needs to identify the different elements of the data and then group them into several different categories based on characteristics of similarities, differences, and other factors.

These characteristics can be any number of things ranging from identifying handwriting samples to the types of documents received. If this were code, the machine could only do one single function but because it is an algorithm which can be altered to fit a wide variety of things, the computer can receive this data and classify all sorts of groups that fit within the specific parameters of the circumstances.

This is how machines can change their functions to adapt to the situation at hand. Your email account can analyze all the emails you received, based on a pattern that you have followed, and it divides them into different groups. It can identify which emails are important and you should see right away, those that are spam and junk mail, and even sort out those that may pose a risk to your computer because it carries a virus or malware.

With these types of algorithms, machines can now learn by observing your habits and patterns and adjust their behavior accordingly. So, the very secret to a successful and effective neural pathway depends a great deal on the algorithms your system uses.

Types of algorithms

Without algorithms, machines cannot learn. So, over the years many different ones have been developed. Depending on what you want your machine to do, they can be grouped into two different categories: supervised and unsupervised.

Supervised

A supervised algorithm requires a detailed input of related data over a period of time. Once all the information is available to the computer, it is used to classify any new data relating to it. The computer then does a series of calculations, comparisons, and analysis before it makes a decision.

This type of algorithm requires an extensive amount of information to be programmed into the system so that the computer can make the right decision. That way, when it needs to solve a problem, it will attempt to determine which mathematical function it needs to use in order to find the correct solution. With the right series of algorithms already programmed into the system, the machine can sift through all types of data in order to find the solution to a wide variety of problems in the related category.

Supervised algorithms are referred that way because they require human input to ensure that the computer has the right data to process the information it receives.

Unsupervised

An unsupervised algorithm implies that the computer does not have all the information to make a decision. Maybe it has some of the data needed but one or two factors may be missing. This is kind of like the algebra problems you encountered in school. You may have two factors in the problem but you must solve the third on your own. $A + b = c$. If you know A but you have no idea what b is then you need to plug the information into an equation to solve the problem.

With unsupervised learning, this can be an extremely complex type of problem to solve. For this type of problem, you'll need an algorithm that recognizes various elements of a problem and can incorporate that into the equation. Another type of algorithm will look for any inconsistencies in the data and try to solve the problem by analyzing those.

Unsupervised algorithms clearly are much more complex than the supervised algorithms. While they may start with some data to solve a problem, they do not have all the information so they must be equipped with the tools to find those missing elements without having a human to provide all the pieces of the puzzle for them.

Aside from the two major types of algorithms, there are a number of other types that might be used to teach a machine to learn.

Reinforcement learning

This type of algorithm allows the system to interact with the environment in an effort to attain a certain goal. Reinforcement learning is commonly used in video games where the computer must navigate and adjust its movements in order to win the game. A reward system is used so the computer knows and understands when it should make the right move, but there are also negative consequences whenever they make errors. This type of algorithm works best in situations where the computer has an obstacle that it must overcome like a rival in a game, or it could also be a self-driving car that needs to reach its destination. The entire focus of the computer is to accomplish certain tasks while navigating the unpredictable environment around it. With each mistake, the computer will readjust its moves in order to reduce the number of errors so it can achieve the desired result.

Semi-supervised learning

Semi-supervised learning is a blend of both supervised and reinforcement learning. The computer is given an incomplete set of data from which to work. Some of the data include specific examples of previous decisions made with the available data while other data is missing completely. These algorithms work on solving a specific problem or performing very specific functions that will help them achieve their goals.

Of course, these are not the only algorithms that can be used in a computer program to help the machine learning. But, the general idea is the same. The algorithm must fit with the problem the computer needs to solve.

With artificial neural networks applying these different 'secret formulas' many computers can perform functions, solve problems, and carry on a vast variety of learning processes they could not be capable of doing. This field of computer programming is termed 'deep learning,' a subset of machine learning that makes up the foundation of the material contained in this book.

Representation of algorithms

In this chapter, we will see an adequate and convenient way to represent our algorithms, which will then be translated into some programming language, in order to be executed by a computer.

This means that now we will open a different toolbox, with which we will be able to represent our algorithms.

What must be clear at this point is that we will present new tools and give some examples of how to use them, without pretending to propose an encyclopedia of solved problems, for which there are other books.

Reinforcing more the ideas, if this were a book that teaches to make paintings on canvas, it would show the different tools and its suitable form to use them, advising on different types of paintings and textures, the advantage of its qualities and how to give a good finish to the created pieces. In the middle of this path, examples will be proposed, but that does not mean that it is intended to teach how to make a painting that proposes a human torso, a face, a horse, or any other imaginable object, but it is intended to teach the correct use of tools in order to the reader can begin to make their own pieces, improving the results obtained over time with practice.

In this way, this chapter will be different from the previous one, focusing more on the creation of solutions that can be translated into a program that works on some device.

So far we've seen different algorithms and somehow their way of representing them. That representation was not always the same, because it has to be appropriate for the subject who has to solve the problem. In the case of cooking recipes, we can say that the presentation of the algorithms is done through a textual story, in natural language of the tasks to be performed; in the case of robots or monkeys to the elementary tasks we assign letters, which they know how to understand.

From here on, we will build algorithms that will be executed by a computer. In order for this to be done, such algorithms will have to be expressed in some language that the given subject (the computer) understands. These are programming languages.

When two people communicate, in order to understand each other, they both have to know the language each other speaks. The easiest way is for them to speak the same language. In order to communicate with the computer, and give it instructions to solve a problem, we need a common code, and that's a programming language. Does that mean we have to express our algorithms through programming languages?

The answer is "not necessarily".

Let's follow this path: the algorithms we will create will be executed by a computer. For this occurs, we must express it in some way that the computer understands it, which is through a programming language, but we need an intermediate step to achieve it. Our algorithm, which is in our mind, must be interpreted by some subject who translates it into a programming language. For this we will use tools that fulfill the role of translators of our ideas to future programs.

Our task in this instance will be to represent our algorithms using a tool that allows their subsequent translation into a programming language, either by ourselves or by a programmer.

The way forward would be as follows:

Problem > Algorithm > **Algorithm representation** > Computer program > Computer running the algorithm.

Up to here we have traveled the path that reaches the representation of the algorithm, but for very special subjects. Now we need to see a new form of expression, so that a programmer can build a program, in some language.

Making analogies, if we want to build a house, we can say that the objective of this chapter is to build the blueprints that allow its construction. On one side are the ideas that determine what the house will be like, and on the other side is the house built. In the middle are the blueprints.

If we want to build a nice and useful house, we can imagine it, but to make it a reality, we first need to have blueprints that represent our ideas so that someone can build it according to our preferences. What we will see in this chapter, will be something like drawing up the blueprints of our ideas.

Strictly speaking, we've already been doing that, but now we'll go to those special persons who are capable of building programs that work on computers, and we won't be dealing with robots or monkeys anymore.

We can be good at solving problems and creating algorithms, but that doesn't make us good programmers, because we should also know the programming languages, which are thousands.

Having these models is important, because they represent, in an understandable and univocal way, the solution to a problem that can be translated into different programming languages. If for some reason it is decided that a finished system should be reprogrammed in another language, having the corresponding models will be of great help.

In short, and with the intention of being repetitive, what we need is a tool that allows us to represent our algorithms in a way that can be understood by a programmer who can translate it into a program in a specific language.

We can carry out this task just as we do with a cooking recipe: we could tell the programmer in natural written language how to solve the problem, and he would then translate that story into a program.

This would not be entirely correct or practical. The problem with this solution is that the natural language is ambiguous, and ambiguities can lead to confusion, and confusion would lead to incorrect programs. That we must avoid, so natural language is discarded as a suitable tool for the representation of the algorithm.

Reinforcing the idea, natural language can be a good tool for representing recipes, but not for specifying computer programs.

There are graphic tools composed of figures and text that together form a diagram of the solution to the problem.

These techniques are known as algorithm diagramming. The most widespread is called flow diagram, which contains figures indicating actions, connected through lines that determine the flow of execution of the algorithm. The problem is that with this kind of diagrams it is possible to build solutions that are unclear and sometimes impossible to program with current languages.

Another form of graphical representation of algorithms is through Nassi-Schneiderman diagrams, which were exposed by their authors in 1973. The advantage of the latter with respect to the flow diagrams, which are earlier, is that they allow you to quickly visualize the structure of the algorithm, clearly showing their control structures.

Pseudocode

Another tool, based on natural language, is pseudocode. These are instructions written in a natural language, but with a reduced set of words, and with a limited number of verbs, which allow ambiguities to be avoided. It is a very useful tool, and very widespread. There are different versions of pseudocode, and we will base ourselves on a particular one that I will define especially for us to understand.

Before presenting the pseudocode that will accompany us in the representation of our algorithms, it is necessary to know the concept of variable.

"A variable is a portion of memory identified by a name, capable of storing data, and its value will remain unchanged until modified."

A variable, in the context of our pseudocode, will be able to store a number, whatever its type, a text, or any simple data. We can make an analogy between a simple data and any data we can use to fill a box in any form.

For didactic purposes, we will work mostly with numerical variables.

Let us look at the different elementary tasks and their corresponding expression in pseudocode.

Before we begin, we will agree on some notation issues:

- Each elementary task will be given the instruction name, and will be represented by one or more words, each of which is called a "keyword".
- The keywords that indicate actions in the algorithms will be written with lower case.
- When describing the syntax of any instruction, we will enclose between <> the text that we should add.
- The variables will be written in capital letters, and can have any number of letters, although generally we will use only one.

Data Entry

enter <list of variables>

The data entry instruction represents the task of asking for data and storing it in a variable. If instead of a variable we put more than one, each data entered will be assigned in the corresponding variable according to its order. If it was a program running on a computer, the execution stops and waits for the operator to enter the required data.

For example:

enter A

Requests a data and its value is stored in the variable A

Let's look at another case:

Requests three data, and they are stored correspondingly in variables A, B and X

Assignation

<variable> = <expression>

This instruction represents the elementary task of assigning the result of a given expression to a variable. This expression can be a variable, a constant, or any combination of variables and constants linked through different operators. The keyword here seems absent, but is actually the sign =. The variable that receives the result of the right expression must always be placed to the left.

The operators we are going to use are the following:

Operator	Description

+	Sum
-	Subtraction
*	Multiplication
/	Division
\	Integer division
**	Power
mod	Rest of integer division

The use of parentheses is allowed to define the order of execution of the operations if necessary.

Perhaps the mod function deserves further explanation:

If we divide two integers, resulting in another integer, the rest of that division is what we will call mod. For example if we divide 25 by 7, the integer result is 3 and the rest of that division is 4. Therefore 25 mod 7 would result in 4.

Examples of assignment instructions:

Instruction	Operation

A=5	Assigns the value 5 to variable A
A=B	Assigns the value of variable B to variable A
C=C+1	Increases the value of variable C by 1
D=X mod 10	Assigns the rest of dividing X by 10 to D.
C=A*(X+Y)	Assigns to C the result of X+Y multiplied by the value of the variable A.

Always the variable that will receive the result of the expression is the one we place on the left side. Thus, in the second example we know that it is A that is being modified by assigning the value of variable B, and not the other way around.

Output

show <list of expressions>

The output instruction is the one that allows to show the results, through the evaluation of expressions. The output can be a constant, a textual message, a mathematical expression such as those used in the allocation blocks, or a combination of all of them.

Examples of output

Instruction	Operation
show A	Displays the value of variable A
show A,B	Displays the values of variables A and B
show S/N	Displays the result of dividing S by N
show "No solution"	Displays the message "No solution"

Let's start with the monkey, and replace it with a computer. We must then build an pseudocode algorithm, which allows a programmer to write a program executable by a computer, in a particular language.

The first problem we solved with the monkey was to add two numbers. As we have already solved the problem, all we have to do is adapt the representation of the algorithm through the pseudocode that we have just presented.

The tasks we must carry out are the following:

-Enter two numbers

-Add them

-Show result

A pseudocode version of the problem, as a representation of the algorithm, would be as follows:

```
enter A
enter B
S = A + B
show S
```

Let's explain what this is all about with an example. Suppose
we want to use this algorithm to add the numbers 5 and 9. The
first instruction is "enter A", which corresponds to our first
data, so what we get is to give variable A the value 5, which
would be one of the data that would have to add. The next
instruction is "enter B", then there we would enter our second
data, which is the number 9, and we assign it to variable B. So
far we have variable A with the value 5 and variable B with the
value 9. The next instruction is "S=A+B", that is, an
assignment operation that places in the variable S the value of
the sum of A and B, in this case the number 14 would be stored
in S. The last instruction is "show S", with what we get to show
the result of S, which is just 14, that is the sum of our data 5
and 9.

Can we solve it like this?

```
enter A,B
S=A+B
show S
```

And so?

```
enter A,B
show A+B
```

Which one's right? All of them!

Which one is better? Depends. The first two are more expressive. The third version is shorter and seems more efficient, but it is not very clear. It is better in these initial instances, to sacrifice a little efficiency in benefit of the clarity of the algorithms.

If we did not try to add two numbers and these were 3, the solution would be very similar, and so would the alternatives. For example:

```
enter A,B,C
D = A+B+C
show D
```

What if there were four of them?

```
enter A,B,C,D
S = A+B+C+D
show S
```

What if they were N numbers?

We still don't have the necessary elements to add a long list of numbers, or a list of a variable number of elements, but soon, in a few pages ahead, we'll be able to do it.

Decisions

Following the same path traveled in the first part of the book, it is time to somehow incorporate the concept of decision to our pseudocode. We know at this point that decisions are important within many of our algorithms, which is why we must have an instruction that represents them in an appropriate way. For this we will use the words "if" and "else". The word "if" will always be followed by a condition, which will have a true or false result. If the true condition is found, then the tasks or instructions we must perform in that case will be executed. On the contrary, if the condition is false, another group of elementary tasks or different instructions will be executed.

```
if <condition> then
    <instructions if the condition is true>
else
    <instructions if the condition is false>
```

Where the condition is expressed through relational and logical operators, which are represented as follows:

Operator	Description

=	Same
>	Greater than
<	Less than
>=	Greater than or equal to
<=	Less than or equal to
<>	Different
^	Logical conjunction or simply AND
v	Logical disjunction or simply O
~	Logic Negation or simply NOT

The instructions, or elementary tasks, corresponding to the true condition may contain **one** or more instructions of any kind.

The instructions, or elementary tasks, corresponding to the false condition may contain **zero** or several instructions of any type. This means that in some cases there may be no "else" clause.

In order to clearly determine the blocks of instructions, the concept of "indentation" is used, which consists of writing the instructions with an alignment to the right, through the use of the insertion of blank spaces or the use of the tabulator. In our case, we will use 4 blank spaces as indentation, to determine the instructions that remain under the effect of the "if" or the "else".

Examples of decisions

"Given two different numbers, show the value of the greater".

The problem is simple to understand: we must show the value of the greater of two numbers that we know are different, therefore we know that one of them is greater. If they were the same, which would be the greater? Possibly the reasonable answer is "none", it can also be "any", but the problem asks to show the value of the greater, so if they were equal would be enough to show the value of any of them. In these cases, in order to avoid philosophizing on matters that are not relevant, it is convenient that the statement clarifies accurately what should be done. To overcome this inconvenience, the statement says "different numbers," and the matter is resolved. Another way to solve it would be to modify the statement: "Given two numbers, show the greater. If they are identical, show the value of any of them". That is how the problems should be enunciated, so that their form of expression is not an additional problem that we have to solve. A statement should always be clear, and should not be ambiguous.

Returning to our problem and following the methodology proposed in the previous chapter, the data or inputs of the problem are two numbers, which we will call A and B.

The expected result is the highest value between A and B.

A model solution, it could be:

If A > B then do G=A, else G=B, and then show the value of G.

It is good to remember that the moment to model a solution is the creative moment of the process. Here we can analyze alternatives, we can disagree, but the important thing is that we arrive at a correct solution. In the case of the proposed solution model, a variable G appears, where I will put the value of the greater between A and B, and then show G.

The algorithm expressed in our pseudocode would look like this:

```
enter A and B
if A > B
    G=A
else
    G=B
show G
```

Algorithm to determine the greater of two different numbers, first version

Another possible solution would be to modify the solution model:

If B>A then do A=B, and then show A

It seems simpler, because it is shorter, the variable G does not appear, but is this solution correct?

This solution proposes that if B is greater than A, then in A we place the value of B, therefore in A the greater will remain, but losing the value that A had before, this is irrelevant to us in this context. If the condition is not met, i.e. B is not greater than A, nothing is done, because just A is greater than B. At the end, A will always have the value of the greater, so I just have to show how is the value of A.

Brilliant? I wouldn't say that much. Perhaps ingenious, thrifty, a miserly algorithm, which stinks of resources at the cost of losing clarity. Anyway, it's right and it can be used.

The corresponding pseudocode would look like this:

```
enter A and B
if B>A
A=B
show A
```

Algorithm to determine the greater of two different numbers, second version

As you can see in the algorithm, if the condition is false, nothing should be done, and on that side there are no instructions, so there is no "else" clause of the condition.

Let's slightly modify the statement: Dice two numbers, show the greatest. If they are the same, display a message that says "they are the same".

The data are the same, but the expected result is different, because a message may appear saying that the numbers are the same.

A model solution could be as follows:

If A=B show "they are the same", but If A>B show A, but show B

The algorithm, in consequence, could look like this:

```
enter A, B
if A=B
    show "are equal"
else
    if A>B
        show A
    else
        show B
```

Algorithm to show the greater of two numbers, being able to be equal, first version

Note the use of indentation to determine the instructions to be executed in each case.

Another possible solution would be:

```
enter A, B
if A=B
    show "are equal"
if A>B
    show A
if B>A
    show B
```

Algorithm to show the greater of two numbers, being able to be equal, second version

Which is better? No doubt the first version, because it is clear and efficient. Ask the fewest questions without losing clarity. Anyway, the elections are subjective.

The second version may ask unnecessary questions. Why? It's an interesting subject to discuss:

Evaluating the first condition, we'd already know if A=B. If this happens, it displays the corresponding message. However, the execution continues, asking unnecessarily if A>B and if B>A, knowing that this will not happen.

On the other hand, if A were not equal to B, but A>B were true, we would show the value of A, but then, equally unnecessarily, we would ask if B>A.

In the latter case, if B were greater than A, we would arrive at the true condition in the third question, which is also unnecessary, because if A is not equal to B or greater, surely B is greater than A.

Obviously this shows that the algorithm is not efficient, but it is still correct.

Here we are dealing with a very simple problem, but the interesting analysis of the proposed solutions could lead us to the roots of artificial intelligence. Is it that bad? Well, the seed, more than the roots. Without going into too much detail, here we have two numbers that are our data, as we ask questions about them we are "learning" characteristics, facts. In the initial solution, we first ask if A is equal to B, if this does not happen, we only ask how A is with respect to B, since we know that they are not equal and there is a major. In the second case, we do not take advantage of the knowledge acquired about the data, and when we ask if A is greater than B, the algorithm does not use the result of the previous question, as if it instantly "forgot" something it should know. Using that idea, we can make algorithms more efficient, and more "intelligent".

Anyway, we insist that in this instance we are not trying to build the most efficient algorithms, but simply the correct ones. However, I will give my opinion on the various solutions in order to justify an "efficient solution" criterion.

Suggested exercise for the reader:

Determine the greater of 3 numbers.

Think of the intelligent version and the "amnesic" version,

which does not take advantage of the knowledge acquired.

Another example of decisions: Sorting playing cards

In the first part of the book, when we were dealing with the concepts of problem solving and algorithms, there was a problem that consisted of ordering 4 playing cards from minor to major. Could we use that algorithm to build a pseudocode algorithm that orders 4 numbers from minor to major? I'm sure you do.

To make things easier, I'm going to transcribe the algorithm from the first part below, just to avoid going back a few pages to find it:

- Locate the 4 disordered cards in the four places A, B, C and D
- If the card in A is greater than the card in B, swap them.
- If the card in A is greater than the card in C, swap them
- If the card in A is greater than the card in D, swap them.
- If the card in B is greater than the card in C, swap them
- If the card in B is greater than the card in D, swap them
- If the card in C is greater than the card in D, swap them

The only problem is that the task "swap" does not exist in our toolbox formed by the given instructions, therefore, we will have to construct a new algorithm to exchange the value of two variables, since it is not an elementary task.

The problem is analogous to exchanging the contents of two cups. How would you do that? You may need a third auxiliary cup to do the following:

- Pour the contents of cup A into the auxiliary cup.
- Pour the contents of cup B into cup A
- Pour the contents of the auxiliary cup into cup B

Problem solved. Depending on the liquids in the glasses, it may have been necessary to rinse them before each step, but luckily the variables do not need to be rinsed.

Done this with variables, and calling the auxiliary as variable X, the solution would be as follows

    X = A
    A = B
    B = X
Algorithm to swap the content of two variables

Translated to an algorithm with 4 variables A, B, C and D, with X as auxiliary, and taking into account the variable swapping algorithm, it would look like this:

    Enter A,B,C,D
    if A>B
        X=A
        A=B
        B=X
    if A>C

```
        X=A
        A=C
        C=X
     if A>D
        X=A
        A=D
        D=X
     if B>C
        X=B
        B=C
        C=X
     if B>D
        X=B
        B=D
        D=X
     if C>D
        X=C
        C=D
        D=X
     Show A,B,C,D
```

Algorithm to order four numbers

A little tedious? We had to ask 6 questions and 6 exchanges.
That's only with four numbers. I estimate that with 5 numbers
it would be 10, with 6 numbers 15, and with 10 numbers 45.
That would cause extremely long algorithms, only for a few
numbers. The reason is that this way of solving the problem
uses more "brute force" than intelligence.

What if we had to order N numbers? There are many
algorithms that solve this problem in a perhaps simpler, or at

least shorter, and certainly smarter way, but we still don't have the necessary elements to build them, and we'll study them later.

As already established in the first part, in order to solve many problems it is necessary to have the possibility of carrying out a set of tasks in a repetitive way, which we will call "cycles".

Requirements of a cycle

The requirements that we must meet when performing repetitive, cyclic or iterative operations are the following:

1. Establishment of some state prior to entering the cycle, through the use of variables whose values determine the beginning of the cycle.

2. Define a condition that determines the execution or completion of the tasks of the cycle, where the variables of point 1 will intervene.

3. The tasks of the cycle, called the body of the cycle.

4. The modification of some element that allows a variable participating in the condition to be modified at any time.

Point 4 is fundamental, because if some value of the variables that is part in the condition of point 2 is not modified, it can happen that the cycle never ends, entering an infinite repetition condition.

Let's see all this within an example already studied, such as beating egg whites to snow point.

Algorithm:

- Putting egg whites in a bowl
- As long as the egg whites are not about to snow repeat the task BEAT

How does this algorithm work? Let's review the fulfillment of the requirements defined above:

1. Putting the egg whites into a bowl establishes the preconditions, initial, essential for the cycle can begin. Without them there's no point in beating.

2. The egg whites did not reach the snow point (condition under which something has to be done). Periodically we will review this condition, which will determine when   we must stop beating, otherwise we will never end.

3. Beating, for example, could be a whisk turn over the

egg whites.

4. The action 3, incorporates air to the egg whites, modifying its consistency, in this way at some point the egg whites, can make condition 2 cease to be fulfilled and the cycle ends. Corresponds to a change in state, or variable, that affects the condition.

Perhaps another example will make it clearer. Suppose we want to eat all the candies in a bag. Will it be clear what the repetitive operation is? Take a candy out of the bag and eat it. When do I finish? When there's no more candies in the bag.

Algorithm:

- Putting candies in a bag

- While there are some candy in the bag, repeat the following: take a candy and eat it.

The explanation of the algorithm:

1. We must have a bag of candies, which sets the preconditions

2. As long as there are candies in the bag, continue (cycle condition).

3. Taking a candy and eating it (action that modifies the state that can affect the condition)

4. Action 3 clearly modifies the conditions, which will allow condition 2 at some point to determine the end of the task.

We will represent the cycles through pseudocode in the following way:

    repeat while <condition>
        < cycle body instructions>

Where <condition> is a relational and/or logical expression, with true or false result, under which the body of the cycle will be executed. As in the decision block, the instructions that make up the body of the cycle can be one or several. Even within such a body there can be another cycle, conditions, or any instruction of any kind and it can be as simple or as complex as necessary.

Examples of the use of cycles

Problem Statement: Show natural numbers from 1 to 10

The data of the problem are represented by the limits of the interval of natural numbers that we must show. In this case 1 and 10

To do this we should take a variable, assign it the value 1, show its value and then increase it, while not exceeding 10. Clearly the repetitive tasks are to show and increase.

Let's see how the algorithm expressed in our pseudocode would look:

```
V=1
repeat while V<10
    show V
    V=V+1
```

Algorithm showing the natural numbers from 1 to 10

Now we can make it a little more interesting by showing the first N natural numbers.

We have a new data, which is the number N, which determines the amount of natural numbers to be displayed.

To do this we should take a variable, assign it the value 1 (first natural number, as in the previous example), show its value and then increase it, as long as it does not exceed the value of N. Again, the repetitive tasks are to show and increase.

```
Enter N
V=1
repeat while V<N
    show V
    V=V+1
```

Algorithm showing the first N natural numbers

With this new tool we can also carry out activities that we could not do before. Adding two numbers was simple, adding 3 numbers too, but what will it be like to add N numbers?

The correct formulation of the problem could be: Given N numbers, get the result of its sum.

In this case, the problem data is:

-The value of N, which represents the number of numbers in the list

-The N numbers

The expected result is the sum of the N numbers

The solution model includes the following elements (keep in mind as always that this is the creative part of the process and there can be many ways to solve the same problem).

1. We first need to know the value of N
2. We must count the number of numbers we enter, so that they are exactly N
3. We add each number entered, one at a time
4. At the end we must show the result of the sum

For point 2 we'll need a variable that counts every number you enter and we'll call it C, which should start at zero. This variable will act as a "counter" of events. Why does it start at zero? Because we are going to count how many elements we have added up to a given moment, and it turns out that before starting we have not added any, therefore the counter C is zero. The name of the variable C is totally arbitrary, we could have called it COUNTER.

For point 3 we will need a variable that adds each number entered and we will call it S, which must also start at zero. This variable will act as an "accumulator" for the numbers in the list. Obviously it starts at zero for the same reason that makes C start at zero.

At the end of point 4, we must show the value of S.

```
Enter N
C=0
S=0
repeat while C<N
    enter X
    S=S+X
    C=C+1
show S
```

Algorithm showing the sum of N given numbers

To better understand how this algorithm works, we are going to perform a "desktop test", which consists of creating a table, where each column represents a variable of the algorithm, to which we will add one more column to represent the output, that is, the results they show. In each row we will place the values of the variables that result in each iteration.

Suppose our list is as follows: (12,5,3,9)

We have 4 numbers in the list, therefore N will take the value 4. The variable X, will take in each iteration the value of each one of the numbers in the list. We know that variables keep their value as long as they are not modified, and when that happens, the previous value is "forgotten".

Considering these elements, the desktop test could turn out like this:

N	C	S	X	Output	Comments
4	0	0			Initial values
	1	12	12		First iteration
	2	17	5		Second iteration
	3	20	3		Third iteration
	4	29	9		Fourth iteration and C reaches the value of N
				29	You exit the cycle, and the result S is displayed.

Desktop test of the sum of N numbers

As we always point out, there are many possible solutions to the same problem. Without trying to confuse the reader, it is possible to solve this problem by using the C counter in a different way. For example, we can make it start at 1, in which case the end condition would be a little different: we will no longer ask C to reach the N value to finish the cycle, but to surpass it, leaving the algorithm as follows:

```
enter N
C=1
S=0
repeat while C<=N
    enter X
    S=S+X
    C=C+1
show S
```

Do we encourage more?

Let's take advantage of what we've seen so far to solve problems that pose more complex challenges.

Suppose we have a natural number, and we want to count how many digits it has. Is this a complex challenge? Apparently for us it is very simple to perform that task, and determine for example that the number 7524 has 4 digits. But what mental and mathematical processes were involved in this challenge? I dare to say that we really don't know them, at least on a conscious level. The human brain has a very complete and complex "toolbox" that allows it to solve some problems like this in an apparently trivial way. The challenge is to build an algorithm with the elements we've seen so far. For that we are going to do an analysis of the problem from a mathematical point of view.

We must never lose sight of the steps necessary to solve a problem, which are the model of the solution. As we progress through the book we may be a little more lax at this point due to the accumulation of practice, but for now we are beginners, so let's specify our models in a detailed way.

Step 1: Understand the problem. Given a natural number, determine the number of digits. There is no difficulty in understanding. We just need to determine and report how many digits a number has.

Step 2: Input data. A natural number, which we know is an integer greater than or equal to 1.

Step 3: Expected result. Number of digits of the natural number.

Step 4: Build a model of the solution. Here I insist with the creative process that each one does in his own way to be able to arrive to the results starting from the data, all of that using the tools that we have, that in this case is not our great surprising brain full of resources, but we only have the operations that we define until here and that the only we can use, putting conditions to the space of the solution.

Within this context, we know that we have variables that can store numbers. We also know that a natural number has an implicit decimal comma at the end, that if we manage to run it to the beginning of the number, eliminating its decimal part and counting how many times we do it, we will obtain the number of digits of the number. This will happen by removing the least significant digit from the number until we reach zero.

Final draft:

- we enter the natural number,
- we somehow got a digit out of him,
- we count in some variable what we just did (that variable should start at zero),
- we repeat this until there are no more digits left in the number, that is, until it is reduced to zero.

For the proposed example, we have the number 7524, leaving the process more or less like this:

- To 7524 we take the digit 4 and count it, getting a digit and the number is now 752
- To 752 we take out the digit 2 and we count it, obtaining already two digits and the number now is 75
- To 75 we take out the digit 5 and we count it, obtaining already three digits and the number now is 7
- To 7 we take out the digit 7 and we count it, obtaining already four digits and the number now is 0
- Since we no longer have digits, the algorithm ends and the counter is four.

Now the question would be where in my toolbox does it say we can get a digit out of a number?

The answer is the integer division operation that we symbolize with the operator "\".

7524\10 is 752, which means that we "pull" the last digit out of the number.

So if we do this as long as the number is greater than zero, counting the number of times we repeat the task, we will solve the problem.

Let's see how the solution we already have would look in our pseudocode:

```
enter N
C=0
repeat while N>0
    N=N\10
    C=C+1
show C
```

Algorithm to count the digits of a natural number

Let's see the corresponding desktop test, for the same example, N=7524

N	C	Departure	Comments
7524	0		We enter the data, and set C to zero.

752	1		First iteration, we remove a digit and count it
75	2		Second iteration, we remove another digit and count it
7	3		Third iteration, we remove another digit and count it.
0	4		Fourth iteration, we remove the last digit and count it.
		4	You exit the cycle, and the result C is displayed.

What if instead of wanting to count the digits we want to add them?

We modify the statement by saying: Given a natural number, determine the sum of its digits.

It's simple again, isn't it? So if we have the number 7524 the expected result is 7+5+2+4=18.

Following the methodology:

Step 1: Problem understood

Step 2: Same as above problem, a natural number

Step 3: The sum of the digits of the natural number

Step 4: It looks a lot like the previous problem, but instead of counting the digits, we should add them. For that it is not enough to "take it out", because we would lose its value. Then we should first get its value, add it up and then discard it. Something like that:

·   Enter a number,
·   Get the last digit,
·   Add it to some variable (that variable should start at zero),
·   Get the last digit out of the number,
·   Repeat until there are no more digits left

The new element here is "get the last digit". For that we have another known mathematical operation that we had called "mod", which determines the value of the rest of the integer division.

If we divide 7524 by 10, the result is 752 as a quotient and 4 as the rest. Just "mod" is the rest of the division, that is, the last digit of the number, if we use as divisor the number 10.

In the case of the example it would be:

·   We get the last digit by doing 7524 mod 10, the result is 4 and we add it (the sum is 4),
·   We get a digit out of it and it's 752,
·   We obtain the last digit doing 752 mod 10, the result is 2 and we add it (the sum gives 6),
·   We get another digit and it's 75.
·   We get the last digit by doing 75 mod 10, the result is 5 and we add it (the sum gives 11),
·   We get another one and it's already 7,
·   We obtain the last digit doing 7 mod 10, the result is 7

and we add it (the sum gives 18),
- We take one more and it is 0, then the cycle ends and 18 is the sum of the digits, which is the expected result.

Let's see how **the solution we already have** would look in our pseudocode:

```
enter N
S=0
repeat while N>0
  D=D mod 10
    S=S+D
    N=10
show S
```

Algorithm for adding the digits of a natural number

We emphasize the text "the solution we already have" because it is important to remember that the pseudocode does not solve the problems, but represents their solution. The process prior to the writing of the pseudocode is the one that really resolves it.

Let's see the corresponding desktop test, for the same example, N=7524

N	S	D	Output	Comments
7524	0			We enter the data, and set S to zero.

752	4	4		First iteration, we remove a digit and add it up
75	6	2		Second iteration, we remove another digit and add it up
7	11	5		Third iteration, we remove another digit and add it up
0	18	7		Fourth iteration, we remove the last digit and add it up.
			18	You exit the cycle, and the result S is displayed.

Cycles driven by a variable

When we have to perform repetitive tasks, we often come across a particular case in which the number of times the body of the cycle must be repeated depends only on the value of a variable that receives fixed increments in each iteration.

Many languages took note of this, and created specific instructions to make the programming of these cases simpler.

It's not really about anything new that we can't solve in the way we already know, but to represent these cases we're going to use a new instruction, or rather, a modification of the "repeat" instruction, and thus support with our pseudocode what most languages includes.

The proposed form is as follows:

> repeat for <variable> = <initial value> increment <increment value> up to <final value>

Both <initial value> and <increment value> and <final value> can be constant values, the value of a variable, or expressions in general.

It is important to emphasize that all the variables that intervene in this sentence must not be modified in the body of the cycle, but will be modified automatically according to the value determined in the argument <increment value>.

The operation is as follows:

1) The variable takes the initial value and verifies that it has not exceeded the final value witch ends the cycle.
2) The body of the cycle is executed
3) The variable is automatically increased by the value of the increment.
4) Back to step 1

The operation described corresponds to cycles with ascending values. If the final value is less than the initial value, the increment should be a negative value. If we want to "count backwards", step 1 would be modified by evaluating that the initial value is not less than the final value.

Examples

Previously we had seen the following algorithm to show the first 10 natural numbers

```
V=1
repeat while V<10
    show V
    V=V+1
```

This is a typical case of a cycle driven by a variable. With our new instruction it would look like this:

```
repeat for V=1 increment 1 to 10
    show V
```

Algorithm to show the first 10 natural numbers with cycles driven by a variable

We can observe all the elements of a cycle seen before, because in the same instruction the initial value, the condition and the variation of the variable are established. The variable V would begin with value 1, which is the <initial value>, would control that this value does not exceed the <initial value>. If this does not happen, it executes the sentences of the body of the cycle, and at the end it increases the variable V by 1, which is the value of parameter <increment value>.

If we wanted to count down from 10 to 1, the algorithm would be as follows:

```
repeat for V=10 increment -1 to 1
   show V
```

Countdown Algorithm

Cleaning up the concept of cycle, it is nothing more than a set of instructions that need to be executed repeatedly to solve a problem. Clearly I would have achieved with a "repeat" instruction, but in order to get closer to what most languages offer, we will use the "repeat while" and "repeat for" instructions. The only difference is that the first is based simply on the value of a proposition, and the second will be used when the *number of times the body of the cycle must be executed is perfectly determined*, either by constant values or by the values of certain variables.

In successive examples we will see some uses of the two forms of the "repeat" instruction, and we will clearly indicate in which cases it is convenient to use one or the other.

Indexed variables

So far we have solved some exercises in which lists of numbers participate, but in no case was it necessary to "memorize" the complete list. To solve problems such as adding the numbers on a list, or counting how many meet a condition or getting the value of the average, it is not necessary to remember every item on the list. It is enough to store each element in the same variable that changes its value successively.

However, there are problems where it is necessary to keep all the items on the list, and this situation is observed at least twice:

1) You need to access items on the list in a different order than they were entered, or in an absolutely random order.

2) You need to access the items in the list several times.

Let us think of a problem that can easily and clearly illustrate the first of the cases.

Given a list of 3 numbers, display it in reverse order.

Without thinking too much, we can build the following algorithm

```
enter A, B, C
show C, B, A
```

That was easy, but it only works for a list of 3 numbers. What if the list had 5 numbers? It's not complicated either

```
enter A, B, C, D, E
show E, D, C, B, A
```

Again we have a simple solution, for a limited problem. If our list doesn't have 5 elements, the algorithm won't work.

What if the list had 329 numbers? Or 75? What if the items involved in each case vary in quantity?

The wording of the problem now would be:

Given a list of N numbers, display it in reverse order.

We need to store N numbers somewhere, and it seems that using successive letters of the alphabet is not possible, because we don't know how many elements we will have to store each time.

For this purpose, there are indexed variables, which are nothing more than variables identified by a name, and a number that works as a sub-index.

If the variable has only one subscript, it is called a vector, and is used to operate with lists.

If the variable has two subscripts, it receives the name of matrix, and is used to operate with tables.

Variables can have more indices, but for the purposes of this book, we will work with variables from a sub-index, because it is enough to illustrate the concept, without going into unnecessary algorithmic complexities.

Let's see, then, the particular case of a variable with only one subscript or dimension, which is enough for us to work with our lists of numbers.

A vector, then we will represent it in the following way:

<variable>[<sub-index>]

Where the subindex is an expression that results in a number that indicates the position within the list or vector. Obviously we are interested in the integer part of the result of this expression, because just as its name indicates, this value represents the position of the element within the vector.

Example of vectors

Suppose we have vector V, which is going to contain 5 elements. Then we would have 5 variables identified as follows:

V[1], V[2], V[3], V[4] and V[5].

Making the analogy with the little boxes that represent the variables, we can say that we have then five little boxes, which are all called V, but each of them is identified with a number.

That number can be given by an expression, for example:

V[5-2] would represent element V[3]

V[H] will depend on the value of the variable H

V[J+K+1] will depend on the values of J and K

As we can see, the sub-index can be any arithmetic expression, from which we will take its integer part to indicate the element of the list to which we are referring. No matter what variables you use within the subindex expression, no matter how complex or simple the expression, the only relevant thing is its value.

Returning to our original problem, if we have a list of N numbers and we want to store it in a vector that we will call L, we can do it with the following algorithm:

```
enter N
repeat for I=1 increment 1 to N
    enter L[I]
```

Algorithm for entering a vector or list size N

The variable N will take the size of the list, when the cycle starts the variable I will take the value 1, so when the instruction is executed enter L[I], actually we will be entering a value for L[1], then I increases and in the second pass through the cycle will be worth 2, then we will enter a value for L[2], and so on, until we have the list completely entered in the vector L.

To solve our problem now we should show the list in reverse

order to the one entered. We must add to the previous algorithm, another cycle that starts with N and ends in 1 for this to happen:

```
enter N
repeat for I=1 increment 1 to N
    enter L[I]
repeat for I=N increment -1 to 1
    show L[I]
```

Algorithm for showing a list in reverse order

It would work the same way if you used one variable to control the first cycle and another for the second, since to identify the elements of a vector only the value of the sub-index expression matters.

```
enter N
repeat for I=1 increment 1 to N
    enter L[I]
repeat for J=N increment -1 to 1
    show L[J]
```

We can also complicate things unnecessarily by solving the same problem in the following way:

```
enter N
repeat for I=1 increment 1 to N
```

```
    enter L[I]
  repeat for I=1 increment 1 to N
    show L[N-I+1]
```

It would work the same way, but with an unnecessarily more complex expression in the vector index.

What is clear is that there is no single solution to a problem, however simple it may be. With time and practice we will know, somehow, how to choose the most efficient and clear solution.

Let us suppose a very simple problem to illustrate the second of the cases of use of indexed variables or vectors, that is, when we need to go through the data in the list more than once:

Given a list of N numbers, determine how many items are equal to the average.

We know that to calculate the average of a list of N numbers, we must add all its elements and divide them by N, obtaining for example a value that we will call P. Then we should go through the list again to count how many elements are equal to P. There is no way to solve this problem with "a single pass" over the list. Nor is it prudent to tell the user of our program to enter twice the list: one to calculate the average and another to count how many elements are equal to that average. If the list had about 5 elements it would be annoying but not very serious, but if it was a list of 1658 numbers, it would not be practical or nice at all.

Consequently, we might think that in order to solve this problem we should:

- Enter the list and store it in memory in some way
- Add your elements
- Calculate the average as the sum obtained divided by the number of elements: N
- Scroll through the list counting how many times the average obtained in it appears, for this you should use a variable that acts as a counter, which will start at zero.

Following this model, we can build the following algorithm

```
enter N
repeat for I=1 increment 1 to N
    enter L[i]
```

So far we have the list entered in the vector L, now we are going to add its elements, in the variable S

```
S=0
repeat for I=1 increment 1 to N
    S=S+L[i]
```

Now that we have the sum of all elements in the list in S, we can calculate the average P

```
P=S/N
```

We already know the average and we can see how many times it appears on the list. We will use the variable C as counter

```
C=0
repeat for I=1 increment 1 to N
    if L[i] = P then
        C=C+1
```

We are already able to show the result

```
show C
```

Putting it all together, the algorithm would look like this:

```
enter N
repeat for I=1 increment 1 to N
    enter L[i]
S=0
repeat for I=1 increment 1 to N
    S=S+L[i]
P=S/N
C=0
repeat for I=1 increment 1 to N
    if L[i] = P then
        C=C+1
show C
```

Algorithm to show how many times the average is in a list

We can try a slightly more efficient solution if we join the first two cycles into one, since we can add the elements of the list as they enter.

```
enter N
S=0
repeat for I=1 increment 1 to N
   enter L[i]
   S=S+L[i]
P=S/N
C=0
repeat for I=1 increment 1 to N
    if L[i] = P then
       C=C+1
show C
```

That's it?
All the tools of interest of this book are already given as far as algorithm representation is concerned. However we are going to see some more typical problems, just for illustration purposes.

Sort a list of N numbers

Near the beginning of this chapter, we had built an algorithm that allowed us to order 4 numbers, and the promise to see what would happen if instead of 4 there were more elements to order was latent. We are now in a position to tackle a solution to that problem, which without vectors would have been impossible to resolve.

There are many list ordering algorithms, some more efficient than others. Here we will show a simple one, with low complexity, although not very efficient but effective. In fact, it is one of the least efficient but also one of the simplest; and since it is an introductory book to problem solving, I consider it more appropriate to focus on this type of algorithms.

What we did almost without wanting to order the 4 numbers, was to make a complete tour of the list by placing in the first place the minor of all, then we did the same with the remaining 3 numbers, and then with the last two, thus being ordered the list.

Let's generalize this solution: to order N numbers from lowest to highest, we must go through the complete list, comparing its elements so as to always place the lowest of all in the first place. Then we repeat the same procedure for the second element, comparing it with the rest of the list and making the exchanges of elements necessary for the minor to remain there. If we repeat this procedure all the remaining elements until the list is exhausted, the list will be sorted. When we analyze the ith element we will always have to the left an ordered sublist, and to the right a disordered sublist.

To do this we must first enter the list of N numbers.

Then, for each of them, from the first to the penultimate, go through the rest of the list always choosing the minor in the appropriate place. Why to the penultimate? Because the rest of the list of the penultimate element, is exactly the last element and therefore the last that I must compare so that the list is ordered.

This seems like a complicated play on words, we can express it in a different way:

- Let's take a cycle on variable I, which runs through the list taking values from 1 to N-1.
- Within this cycle we will include another one, for which we will use the variable J, that goes through the list from I+1 to N, which is no more than the rest of the list for each value that I takes.
- If the element in position J is smaller than the one in position I, then they must be exchanged.
- Then, we can show the list, which was ranked from minor to major.

Expressed in a more mathematical language we can say:

Given a list L of N numbers, for all I between 1 and N-1 and for all J between I+1 and N, if L[J]<L[I], then swap them.

Is admirable the simplicity of mathematical language. Now let's build the algorithm, taking into account that at the end we must show the list as it was, that is, ordered.

```
enter N
repeat for I=1 increment 1 to N
    enter L[I]
repeat for I=1 increment 1 to N-1
    repeat for J=I+1 increment 1 to N
        if L[J]<L[I] then
            X=L[J]
            L[J]=L[I]
            L[I]=X
```

```
repeat for I=1 increment 1 to N
    show L[I]
```

Algorithm to sort a list of N numbers from lowest to highest

Intersection of sets

Suppose we have two lists of M and N elements respectively representing sets A and B. Since we suppose that they are sets, they will not have repeated elements, but they will not be ordered either. We must obtain a set C, which is the result of the intersection of A and B, that is to say that each element of C must be in A and in B.

How do we solve this problem?

We will represent each set with a vector. This way of doing it is a license that we take, because in mathematical rigor they are totally different things.

In the case of the intersection we can, for example, take each element from A, and if it is also in B, we will add it to C.

Let's solve the problem step by step:

The data of our problem are two sets A and B, of M and N elements respectively, which we will represent through two vectors, which we know will not have repeated elements.

We must obtain a set C, with the common elements of A and B.

- A model of the solution could be like this:
- Enter set A
- Enter set B
- For each element of A, if it is in B, add it to the set C
- Show set C

To execute the fourth step, we must go through all the elements of A, and for each of them we must go through B, but do we have to go through all the vector B? Only as long as it doesn't find the element of A in B, or as long as it has elements in B. So there's a double condition there.

The cycle on A will be one of the "repeat for" type, because it must be executed for all the elements of A, on the other hand the cycle through vector B will be a cycle of the "repeat while" type with the double condition mentioned above. When we leave the cycle that goes through B we must **ask what happened**. We leave because we find the element, and consequently add it to the set C, or because the vector is finished, in which case the element of A is not in B and we must discard it. Let's see how we express these ideas in our pseudocode, but now interspersing separate comments between asterisks *:

```
*First we enter the set A of size M
enter M
repeat for I=1 increment 1 to M
    enter A[I]
*Then we enter the set B of size N
enter N
repeat for J=1 increment 1 to N
    enter B[J]
*We put in zero a K counter, that we will use to know how
```

many elements will have the set C*.
K=0
*Now we'll go through all the elements in set A with one
cycle.
repeat for I=1 increment 1 to M
   *We will search if the I element of A is in B
   J=1
   repeat while (J<=N) ^(A[I]<>B[J])
     J=J+1
   *Now that we're out of the search cycle, we ask what
happened
   *If we find the element of A in B, we add it to C
   if A[I]=B[J]
     K=K+1
     C[J]
*Once we're done, we show set C
repeat for I=1 increment 1 to K
   show C[I]

Now let's see the same algorithm without the comments:

enter M
repeat for I=1 increment 1 to M
   enter A[I]
enter N
repeat for J=1 increment 1 to N
   enter B[J]
K=0
repeat for I=1 increment 1 to M
   J=1
   repeat while (J<=N) ^(A[I]<>B[J])
     J=J+1
   if A[I]=B[J]

```
     K=K+1
     C[J]
repeat for I=1 increment 1 to K
     show C[I]
```

Deep Learning

The innovations and breakthroughs that have been accomplished with deep learning are enough to get a lot of people excited about it. It is said that we live in the age of technology and with the many strides already achieved by deep learning, the next age will probably be the age of artificial intelligence. To that end, there will be more and more people looking to enter this field, so they can have a role in ushering in this new era.

However, before they can do that there are a lot of things you must come to understand. Now that these fundamentals are so clearly realized, we will begin to see artificial intelligence emerge in all sorts of industries. We've already discussed how it is being used in areas of pattern recognition, and object detection, but the extent of deep learning will one day go much further than that.

This is a complex field and for anyone to get a firm grasp on it, it helps to have a basic understanding of subjects like linear algebra, calculus, probability, and programming. This knowledge will be very beneficial in helping you to grasp the concepts we will be mentioning later in this book. If you don't have this background, it doesn't mean that you cannot grasp these concepts, but it does mean you'll have to do a little extra homework to get through the meat of it all.

You'll also need to know some practical differences in how deep learning differs and why it is so much more efficient than any other type of artificial intelligence before it.

Deep Learning, Machine Learning, Artificial Intelligence
– What's the Difference?

Often when you research deep learning you'll come across several different terms that seem to be used interchangeably. Deep learning, machine learning, and artificial intelligence. While these are all related there are some distinct differences that set them all apart from each other.

So, what exactly is deep learning? Basically, it is a type of machine learning that uses a system of neural networks that are designed to mimic the learning of the human brain. This network's focus is to simplify how a machine uses certain learning algorithms, which can be applied in artificial intelligence and machine learning. In fact, it is virtually impossible to have artificial intelligence without deep learning.

Because of deep learning, we now have the ability to develop massively large neural networks that have the ability to absorb information and respond to it without the need for a human to program it's every move. The expectation is that one day, these machines will be able to operate autonomously, completely on their own without any form of human interference.

For the layperson, we are familiar with the term "artificial intelligence." It is the stuff of countless science fiction movies. However, few people understand that it is no longer science fiction but is definitely emerging in the realm of reality.

To explain:

- Deep learning is a form of machine learning that is used to teach machines how to learn in a way that is similar to the learning process in the human brain.
- Machine learning is a series of algorithms that give machines the ability to collect data, analyze it, and make decisions based on that analysis.
- Artificial intelligence is a machine programmed with machine learning, so it can gather data and make decisions based on that data.

In its most basic of forms, deep learning is simply another type of machine learning. Basic machine learning has improved over the years and deep learning is the latest link in the evolutionary chain. In its early stages of development, deep learning machines were only capable of learning from data that had already been labeled and stored into the machine. This meant it could only function with human supervision. Unsupervised learning was still far into the future, but many believed it was a very real possibility. Now, machines can learn in all sorts of ways.

How a Machine Learns

When it comes to learning, we humans do it without any thinking or programming. Our minds have never had to be trained in how to learn something. From the time we're born, our five senses go immediately into action, collecting data and submitting it to our brains. Our brains process it, analyze it, and then make decisions based on it. If we're walking out at night, our brain automatically assesses the scene, determines a level of danger, and tells our feet to move faster. If we're watching a movie, our brain takes all that information in and determines whether we should cry, laugh, or be angry. We have what is called unsupervised learning. No one has to program those details and responses into us.

In addition, once we learn a concept, it then becomes a foundation from which we can build on. We first learn the alphabet (the initial concept) then we learn the sounds related to that alphabet, then we learn to string sounds together, creating words. Eventually, we'll be able to read and this is called layered learning. With that thought in mind, we can continue to build our knowledge base and grow from there.

The concept behind machine learning is similar. While it is a far cry from doing the amazing things that a human brain can do, its function is based on a similar set of situations. A massive set of neural networks are created, each one communicating with the other. It also collects data and responds to the information it receives.

With artificial intelligence, the machine will be able to take in raw data and respond with a selection of pre-programmed responses. They will be able to "learn" from the data input and build on it. The major difference is that at present, machines can only learn from human input and develop their own concepts from the information gathered. This is one step closer to mimicking the human learning process.

Deep learning is basically a unique form of machine learning that is much more flexible than other previous forms used. By using a nested hierarchy of concepts, it can access many layers of non-linear information processing and achieve both supervised and unsupervised learning, which is ideal for pattern analysis and classification.

Deep neural networks consist of several layers called a hierarchical neural network. In this type of network, every layer has the ability to change its input data into something that is more abstract. The output layer will then combine the various features of input data and formulate a prediction. This method improves calculations, so it is much easier to understand.

While there are several different ways that machines can learn information, their strategies can be categorized into two separate classes. The first being supervised learning that uses labeled data to classify information. This form of learning is based on data that produces "expected answers." For example:

•   Visual Recognition

Imagine an AI designed to identify pedestrians walking across a street. It can be trained by inputting millions of short videos of street scenes collected. Some of the videos will have pedestrians walking while other videos will not. Some of the videos will have many people walking while some may have only one.

With a number of learning algorithms applied to the data, each giving the machine access to the correct answers a variety of models are designed to teach the machine how to identify pedestrians in fast-moving scenes. The algorithms are tested against an unlabeled set of data that will check for accuracy.

- Predictions

Supervised learning can also be used in making different predictions. A machine can be taught to estimate risk by inputting a large number of actual trades made by real investors and the results they received. It can then be asked to give an estimate of risk for each trade based on several fundamental factors of previous trades: price, volume, company, etc.

It then takes its estimated risk and compares it to the historical results during several different time intervals (day, week, month, and year) to determine if its predictions are accurate or within normal expectations.

Unsupervised learning works a little differently. The machine receives input without any related yield factors. The answers are derived purely from the calculations made. The goal of this type of learning is to demonstrate the basic structure or dissemination of the information so that the machine can gather even more details about the information. Unlike with supervised learning, the machine is not given any right answer and there is no human instructor guiding its data. The AI may collect the data and sort it according to its similarities or differences even if there is no classification. The goal is to show a fundamental structure in the information in an effort to get a better understanding of it.

It is referred to as unsupervised learning because there is no instructor guiding it to the right conclusions. The machine will perform its own calculations in an effort to determine the nature of the data that it has collected.

While there are obvious advantages to unsupervised learning, there are still some problems it has yet to resolve. For example, a machine may be capable of identifying basic visual images (it can tell a cat from a dog) it may also end up creating new classifications in an attempt to distinguish between varying differences within a certain classification. (It may not be able to tell a German shepherd from a Chihuahua). Its purpose is to find relationships within the data it receives but this can create several problems when it tries to go further than that.

Clustering is when the AI attempts to decipher different groups within the data whereas association is an attempt to determine specific rules that will describe the data. Both of these can present huge problems in AI when their results are skewed by unanticipated groupings.

In reinforcement learning, the machine is trained to recognize activities. To come to this conclusion, it learns from its own actions and not from a human instructor giving it the necessary input. The goal, in this case, is not to make a classification or a prediction of events but instead to develop a policy of behavior.

We can find a perfect example of this in our relationship with household pets. If you're going to teach your dog a new trick, you can't just input the information and give it instructions on how you want it to behave. However, if you reward it for doing something and penalize it for other actions, eventually the dog will learn how to do the things that give it rewards and avoid the behavior and those actions that bring on discipline.

Reinforcement learning in machines works in a similar way with a few differences.

- Substitute the pet with the machine
- Substitute the treat for a reward function
- Substitute the good behavior with a resultant action

In order for this to work, you need to have a feedback loop that will reinforce what the machine is actually learning. It is rewarded when it performs certain actions and is disciplined when it is wrong. You might wonder how you can reward or discipline a machine that has no feelings and no emotions. The system will work something like this:

- The machine is given an internal state that it must maintain. This state is used to learn about their environment.
- The reward function is used to teach the machine how to behave.
- The environment is the situation or scenario that the machine must operate in. It consists of all the things the agent can observe and respond to.
- The action is the behavior of the machine.
- The agent performs all the deeds.

Let's apply this to the computer game, Mario. As the machine attempts to play the game it has an environment that allows it to perform many different functions. It does not know what will happen when it performs each of these functions. It can't see the entire environment at a single time, so it must navigate through the environment and make decisions on what to do. If it makes a move that will not advance it, then it is "punished" by not allowing it to move further. If it makes a move that advances it through the environment, it is "rewarded." In time the machine will learn exactly how to navigate safely through its environment until it reaches the conclusion of the game.

This is a very basic explanation of reinforcement learning but it should be enough to give you a general idea of how it works in machine learning. This form of unsupervised learning is not completely without human input. Someone has to create the environment that the machine will operate in as well as the consequences of each move. However, many are looking at this type of machine learning as the true future of artificial intelligence.

There are many applications and uses for machine learning and deep learning by extension that go far beyond the obvious. Already these learning mechanisms are being used in anomaly detection, human genome projects, sequencing analysis, crime analysis, and climatology among countless other uses and no doubt, there will be more practical applications that will be discovered in the future.

How Deep Learning Can Be Used

With the growing emergence of artificial intelligence and deep learning, there is more than enough opportunities for this science to grow and expand beyond its present boundaries. There are three noteworthy events that have been instrumental in projecting this type of technology forward and into the world's consciousness.

The ability of machines to learn and be trained is an important significance for our future. As a result, the framework that has begun to overtake the traditional and outdated technologies in various fields has made it possible for humans to take immense strides in their progress. This can be great news for some people and may cause trouble for others, but the reality is that this technology is here to stay.

Still, like every other modern advancement in the world, people and businesses alike are now learning how to use deep learning to solve a host of real-world problems. However, there are fundamental elements in how deep learning is used and applied in these situations that cannot be ignored.

Pre-training

In machine learning, the process does not focus on collecting numerous datasets. Rather, the machines do the exact opposite. When comparing deep learning techniques to other methods it is important to establish a consistent measure that determines which strategy works better on the same or similar evaluation period. A general rule of thumb is to measure the performance of each strategy based on a set number of datasets using a regular evaluation period.

The problem with this is that in real-world situations, the result is not about how to get an extra percentage out of the error rate but is more focused on building a better robot so to speak. This means that labeling training strategies and highlighting which algorithm used can help the machine to learn better.

To solve many real-world problems, this can turn out to be a very expensive process.

For example, in the field of medicine, a machine designed to detect lymph nodes in the human body by analyzing tomography images (or CT scans) is already in use. This is an extremely time-consuming task because the machine must recognize very small structures. It can also be very expensive as well. Based on the assumption that a radiologist earns around $100/hr and a CT scan can only produce 4 images an hour the cost of such a test could easily run up to $10,000 to get enough images for a proper diagnosis.

Add to that the consultation fee for having an additional doctor on hand to confirm the diagnosis, acquiring sufficient data to give an accurate diagnosis could easily go beyond the quarter of a million mark. That is just for diagnosis only; it does not include any treatment options that will come later.

Credit History

Deep learning is also used in determining your credit score. Machines can learn to analyze patterns to determine who has the highest risk of defaulting on their loan before credit is issued to them. Companies that issue loans to anyone are at the highest risk of finding someone who will default on their loan. This makes issuing credit a very expensive venture. Machines that can learn to analyze spending patterns, payment history, and financial health can make sure that those risks are greatly reduced.

In Computer Games

You've probably already heard about computers that have learned how to play chess or other games. In this type of learning, the pixels on the screen form the basis of the game. The goal and the most complicated task the machine must do is to break through the Deep Mind.

Depending on how complex the game is there are many elements to game playing that the computer has to learn. It will have to navigate through environments, different storylines, character behavior, and other rules in order to master the game.

With it comes to pre-training, the fine-tuning segment learns how to quantify different classes and make the necessary adjustments. Neural networks are pre-trained based on specified datasets and then are fine-tuned to fit within the parameters of a unique problem. Each problem has its own set of different anomalies. The input data informs the machine exactly which layer needs to be adjusted and the learning rates are reset, usually a little higher than the last layer.

In Education

The focus of education is to reach a point where you outperform learning models regardless of the model you choose. When functioning with real-world applications it is not always easy to design a model that functions as it should. It is important that when a learning machine makes an error that it can understand how and why it did so. It must have some ability to ascertain why a specific model did better than any other previous solution, and it is extremely important that you understand that the existing model in use cannot be tricked.

In the Movie Industry

Machine learning is now capable of adding sounds to silent video. The framework was designed to analyze 1000 examples of video playing with a drumstick beating on different surfaces to create different sounds. The machine then studies the video and compares it to a database of pre-recorded sounds and matches the right sound with the scene in the video. The finished scene was then tested for accuracy with humans who were asked to determine which video was the real silent film and which one was matched with sounds added in by computer.

Automatic Language Translation

This type of deep learning trains machines to identify spoken words, expressions, and sentences in an input language and then translate it into a target language. Automatic translation by machines is not a new application but deep learning has been able to add a whole new element to machine translating. It can now automatically translate written text both in printed form and handwritten form and it can also translate images.

Text translation is performed without the use of preprocessing, allowing the algorithms the freedom to learn from the interdependencies that exist between words and then mapping them to the target language.

This function is usually done with the convolutional neural networks because of their ability to identify and recognize all sorts of images. These machines can recognize letters in text, translate them, and then immediately send the translated text to its destination.

Images can also be translated by classifying objects in a photograph as a single set of known objects stored in its memory. Object detection includes being able to identify one or more objects in a photograph and drawing and labeling a box around them.

Handwriting Generation

Probably one of the most impressive applications of deep learning is the ability machines now have to produce handwriting. The machine learns by analyzing a collection of handwriting samples and then produces its own handwritten word or phrase. This is done by inputting a series of coordinates using a digital pen to create its own handwriting samples.

Once the machine has learned the images of samples, it studies the relationship that exists between the pen and the letters to create a new set of handwritten samples to choose from. This may not seem very amazing to you, but the machine can actually learn different handwriting styles and then mimic them back to you.

There are many more uses for deep learning that are in play today. Without your realizing it, you're coming in direct contact with deep learning machines every day. When you turn on your TV, when you make a purchase, use your credit, or even when you go out to eat. Deep learning has already permeated every part of human life and is poised to take us into a whole new era for the world. Ever wonder what deep learning has in store for the future?

What Is It Used for and How?

There is a great deal of excitement surrounding this new technology. With each new advancement, machine learning is gradually moving closer towards handling the more abstract tasks that up until now, have only been successfully done by humans. It is difficult to convince those who are not familiar with the science of the many tasks that machine learning can accomplish.

The different ways this can be used are endless. In fact, there are probably quite a few applications that are affecting your life right now that you didn't realize were being implemented.

Deep learning
Even without a background in computer science, one can understand what the term 'machine learning is.' Basically, it is a machine that learns from data. As long as the machine has the right data input there is a huge number of problems that it can solve without any human interference. As long as the machine is given the correct training data and the right algorithms, it can perform the necessary functions and continue to learn from them for an indefinite period of time.

The primary tools at the heart of deep learning are the input data and algorithms. Without the correct data, it is not possible for deep learning to take place. For years, machines have functioned without algorithms but these machines are programmed to perform certain functions without change (think vending machines) which means the program it started out with will not adapt over time. It will still perform the same actions over and over again regardless of the environment that surrounds it.

However, with deep learning, the computer is capable of making continuous adjustments in order to improve its performance every time it is used. But to really grasp the concept of deep learning, we need to take some time and look closely under the hood. Its use today can help us to see how deep learning is already changing our lives for the better.

When we turn on our home PCs we automatically expect things to happen, and we expect them to happen at record speeds without even thinking about what goes into getting the results we want.

When you turn on Netflix, you will quickly see a list of movies, documentaries, and TV shows that you like. You make your choice without thinking about how deep learning machines have studied your choices over the years and came up with options that are more appealing to you. If you like science fiction, miraculously, you will have a host of science fiction options laid out before you. Because the machine has analyzed your entertainment choices you have the best options in front of you to choose from.

Deep learning is also used by Google, to predict what websites you will most likely want to visit. Google's voice and image recognition algorithms are being used in a host of new industries. MIT is now using deep learning to predict events that are most likely to happen in the future. Everywhere you look, deep learning is beginning to permeate all types of industries and expectations are that it will continue to grow in the future.

But for those who don't fully grasp computer science concepts, the thought of deep learning might instill fear instead of excitement. After all, in the past few decades, the subject has been approached with a lot of skepticism and doubt. Movies have portrayed power hungry machines bent on taking over the world, and news reports about every self-functioning machine failure have been exploited to the highest level. It leads some people to believe that machines capable of learning are more of a threat to humanity than a help. So, what exactly is this kind of technology used for and do we really have anything to worry about?

Classification
Deep learning machines have extremely comprehensive databases and sophisticated networks that allow for easy classification of many different forms of data. We might assume that looking at a picture and identifying its contents is pretty basic stuff. To the human eye, images can be classified in a fraction of a second but to a machine, which only sees things in terms of math, they contain many different elements that must first be sorted out.

Deep learning, however, makes it possible for machines to classify any type of data including video, speech, audio, or handwriting and analyze it to come up with a conclusion that would be similar to that of most humans.

Imagine a computer system that can automatically create a record of the number of vehicles that pass through a certain point on a public road during a set time frame. The steps needed for this to happen are immense. Not only would it have to hold a huge database of different types of cars, their shapes, and sizes but it must also capable of processing the data and analyzing it to come up with an acceptable answer.

Comparing the data it receives through its sensors to the data it has stored in the database, it can classify the answer with a pretty high level of accuracy. While humans could easily identify cars by make and model, the idea of having a human standing on a street corner counting and labeling cars would virtually be impossible to achieve. Even if someone could position themselves to count, humans get tired and need to have frequent breaks. They cannot function continuously without stopping. The level of accuracy would be much lower. Yet, automobile manufacturers, government agencies, and other industries could find the information extremely valuable in making decisions for their business.

But deep learning goes even further than this. While the system may already be programmed with a massive database, as the machine operates it will learn even more and increase its knowledge from its experiences. By being programmed to train itself, continuous human interaction is not necessary. The machine will learn from its mistakes in much the same way as humans do.

Pattern recognition
Pattern recognition is probably the oldest form of machine learning but is also the most fundamental. As of 2015, pattern recognition was one of the most popular areas of interest in research labs around the globe. By giving a machine the ability to recognize a character or some other object, the potential for machine learning increased exponentially.

The ability of a machine to recognize handwritten numbers and letters opens the door to a myriad of uses. This ability has been successful in providing insights into the complex movements of the environment, weather changes, and even in the world of finance. But deep learning involves more than just identifying similar characteristics and differences in images. Pattern recognition allows them to draw their own conclusions in regards to the images or videos they are analyzing and tagging them appropriately. Every time they perform this type of analysis, the better it will become at identifying similar situations and unusual anomalies that could affect the outcome.

Right now, the New York City Department of Transportation has joined up with IntelliScape.io to use this technology and get a better understanding of traffic in their area. They can now see patterns in the weather, identify areas where parking violations are more likely to occur, and as a result, inform local officials of these patterns so they can be prepared to respond accordingly.

There are many uses for pattern recognition in many areas. It can be used to expand the 'Internet of Things' by collecting data from any device that is connected to the internet. Cities will use it to understand how people navigate through their streets, urban planners can make better decisions about the location of physical infrastructures and even in the area of conservation, it can be helpful. Instead of using manpower to go out and count trees, drones can be deployed to analyze the number of trees and their health in any given area.

Prediction

The use of predictions can also be used by many industries. Whether it is in the field of medicine to detect abnormal genes in an unborn child or predicting the change in weather, because of the ability of these machines to paint a realistic picture of future possibilities their potential is huge.

Industries are already using this prediction technology in a vast number of fields.

The pharmaceutical industry uses it to determine the exact set of compounds needed to treat a specific disease. They can now predict which medicines will be more effective and use that data to develop new drugs to fight disease. They can also use it to identify alternative treatments that may also be effective.

In the area of cybersecurity, programs like Deep Instinct focuses on predicting where cyber hackers and other online threats may occur so they can develop ways of protecting the end user before an attack actually happens.

In agriculture, crop outputs can be determined even before planting begins. The computer will analyze weather predictions, soil conditions, and quality of seeds to determine how successful a crop will be and how much profit can be gained.

The same can be said in areas of retail, insurance, finance, and even in aerospace. The potential for this type of technology is vast and positive. As more and more industries begin to adopt deep learning and incorporate it into their business strategies, all sorts of events will become much more efficient and accurate.

Years ago, this type of technology was thought to be in the realm of science fiction and fantasy, but today it is a reality. While computers are a long way from the kind of independent thinking and autonomy that could rule the world, they definitely have come a long way from the vending machine era. The future of neural networks and deep learning are extremely progressive and definitely something we can all look forward to with high anticipation.

Deep Learning Applications

As you have learned so far, deep learning is changing how everybody looks at technology. A lot of excitement swirls around artificial intelligence as well as its branches of deep learning and machine learning. With the huge computational power that machines have, they are now able to translate speech and recognize objects in real time. Finally, artificial intelligence is getting smart.

It is believed that there are many deep learning applications that will affect your life in the very near future. In fact, they are probably already making a huge impact. In just the next five to ten years, deep learning development languages, tools, and libraries will end up being the standard components of all software development toolkits.

Let's look at some of the top deep learning applications that will end up ruling our world in 2018 and beyond.

Self-Driving Cars

Companies that work to build driver assistance services for cars, and full-blown self-driving cars just like Google's, have to teach the computer system how to use all, or at least, the key parts of driving by using digital sensor system instead of needing a human's sense. In order to do this, companies will have to start by training algorithms to use a lot of data. This can be looked at as a child learning through replication and experiences. All of these services could end up providing some unexpected business models for several companies.

Healthcare

Skin or breast cancer diagonostics? Monitoring and mobile apps? Maybe a personalized and predictive medicine on the basis of Biobank data? Artificial intelligence is reshaping healthcare, life sciences, and medicine as an industry. AI type innovations are advancing the future of population health and precision medicine management in ways that nobody would have ever believed. Computer-aided diagnosis, decision support tools, quantitative imaging, and computer-aided detection will all play very large roles in the future.

Voice-Activated Assistants and Voice Search

This is probably one the most popular uses for deep learning. All of the big tech giants have made large investments in this area. You can find voice-activated assistants on almost every smartphone. Siri has been available for use since October 2011. The assistant for Android, Google Now, was launched just a year after Siri. Microsoft has introduced the newest assistant in the form of Cortana.

Automatically Placing Audio in Silent Movies

When it comes to this, the system synthesizes the sounds that are similar to the silent movies. This system was trained with a thousand examples from different videos with sounds of a drumstick hitting different types of surfaces and coming up with different types of sounds. Deep learning models associate the frames of the video with a pre-recorded sound database so that it can choose a sound to play and matches up the best with the things going on in the scene.

They use a Turing Test to evaluate the system such as a setup where humans will have to figure out if the video has real or fake sounds. This uses applications of LSTM as well as RNN.

Automatic Machine Translation

This process is where a given word, sentence, or phrase is said in one language and then automatically translated to another language. This technology has been around for a while, but deep learning has gotten the best results in two areas:

Image translations
Text translations

These text translations can be done without the need for pre-processing the sequence, which allows the algorithm to be able to learn the dependencies between the word and the new language mapping.

Automatic Text Generation

This task is one of the most interesting. This is where a body of text has been learned, and new text is created either character-by-character or word-by-word. This model can learn how to capture text styles, forms of sentences, punctuations, and spelling in the body. Large recurrent neural networks are helpful when it comes to learning the relationship between different items in an input string sequence, and it will then generate text.

If you want to learn more about this or to see some other applications, you can check out Andrej Karpathy's blog. He posts a lot about automatic text generation in terms of:

Baby names
Linux source code
Algebraic geometry
Wikipedia articles
Shakespeare
Paul Graham essays

Automatic Handwriting Generation

This task has provided a corpus of examples of handwriting and generates new handwriting for a certain phrase or word. The handwriting is given as coordinate sequences used by a pen once the samples have been created. From the body, the connection of the letters and the pen movement is learned and the new examples are able to be created ad hoc.

Internet Search

Chances are when you hear the word search; your first thought is Google. But there are actually several other search engines out there such as duckduckgo, AOL, Ask, Bing, and Yahoo. Every search engine out there uses some form of a data science algorithm to provide their users the best results for their search query in less than a second. Think about this. Google process over 20 petabytes of data every single day. If there wasn't any data science, Google would not be as good as it is today.

Image Recognition

Another big area of use for deep learning is with image recognition. This tool is used to identify and recognize objects and people in images and to better understand the context and content. This tool has already been used in many sectors such as tourism, retail, social media, gaming, and so on.

The task will require the objects' classification that is in a certain picture as one of a set of objects that it already knew. A complex version of this would be object detection which involves identifying more than one object in a scene of photo and placing a box around it.

Automatic Image Caption Generation

This task is where a certain image is provided and the system has to come up with a caption that describes what is in the photo. In 2014, a boom of deep learning algorithms achieved pretty big results when it came to this problem. It leveraged the work from top models in order to classify and detect objects in pictures.

After an object has been detected in a photo and it has

generated the labels for the object, you will be able to see that the following step would be to change those labels into a coherent descriptive sentence.

Typically, this system will involve using large convolutional neural networks in order to detect the object in a photo and will then use a RNN, such as a LSTM, to change the label into something coherent.

Automatic Colorization

This is the process of adding color to photos that were originally black and white. Deep learning is able to use the objects and the content of the photo to color these images, a lot like how a human operator would approach something like this. The capability leveraged the large convolutional neural networks and great quality that is created for ImageNet and co-opted to help solve the issue of this task. Typically, this approach will mean that there are a large convolutional neural network and many layers that will provide you with the colored image.

This was traditionally performed by hand by humans because of the difficulty of the task.

Advertising

Advertising, another big area that has been changed by the advent of deep learning, has been used by advertisers and publishers to up the relevancy of ads and to boost their ROI of their campaigns. For example, deep learning helps publishers and ad networks to leverage the content so that they can create precisely targeted display advertising, real-time bidding for their ads, data-driven predictive advertising, and many more.

Recommender Systems

Think about the suggestions Amazon gives you. They help you find relevant products from billions of others, but that also improve your experience. There are a lot of companies out there that use this system to promote suggestions that align with their user's internet. The giants of the internet like IMDB, LinkedIn, Netflix, Google Play, Twitter, Amazon, and several more use this type of system to make their user's experience better. The recommendations you see are based upon your previous searches.

Predicting Earthquakes

There was a Harvard scientist that figured out how to use deep learning to teach a computer system to perform viscoelastic computations. These are the computations that are used to predict earthquakes. Until they figured this out, these types of computations were computer intensive, but the deep learning application helped improve calculations by 50,000%. When we are talking about earthquake calculation, timing plays a large and important role. This improvement may just be able to save a life.

Neural Networks for Brain Cancer Detection

A French research team found that finding invasive brain cancer cells while in surgery was hard, mainly because of the lighting in the OR. They discovered that when they used neural networks along with Raman spectroscopy during surgery, it allowed them to detect the cancer cells more easily and lowered the leftover cancer. Actually, this is only a single piece of many over the last couple of months that have matched the workings of advanced classification and recognition with several kinds of cancers and screening tools.

Neural Networks in Finances

Futures markets have been extremely successful since they were created in both developing and developed countries over the last few decades. The reason for it succeeding is due to the leverage futures provide for people who are participants in the market. They examined the trading strategy, which did better because of the leverage by using cost-of-carry relationship and CAPM. The team would then apply the technical trading rules that had been created from spot market prices, on futures market prices that used a hedge ratio based on CAPM. The historical price data of 20 stocks from all of the 10 markets are a part of the analysis.

Automatic Game Playing

This task involves a model of learning how to play a computer-based game using only the pixels that are on the screen. This is a pretty hard task in the realm of deep reinforcement models, which has also been a breakthrough for DeepMind, which was part of Google. Google DeepMind's AlphaGo has expanded and culminated in this.

Activision-Blizzard, Nintendo, Sony, Zynga, and EA Sports have been the leaders in the gaming world and brought it to the next level through data science. Games are now being created by using machine learning algorithms which are able to upgrade and improve playing as the player moves through the game. When you are playing a motion game, the computer analyzes the previous moves to change the way the game performs.

Deep Learning and the Future

While deep learning is slowly making inroads into our everyday lives, there is much more of it in our future to look forward to. It is a rapidly growing form of technology and as a result, we can realistically expect to see much more of it in the years to come. It will continue to be integrated into ways we are familiar with but more importantly, in ways that we may not have yet thought of.

Considering what we already know about neural networks and the steady stream of research projects popping up all across the globe, we don't really need a machine that can learn to predict what the future holds. Perhaps everything we expect may not pan out, but we can be certain that whatever it is, someone, somewhere is already working on it. Let's take a look at some of the predictions others have made about the future of deep learning to see where we're going.

A Function for the General Population

Some expect that one-day deep learning and artificial intelligence will be available for the general population in much the same way as the personal computer is now in nearly every household. Already, people are working on using this technology to build smart homes that will anticipate your every need, self-driving cars are just one generation away, and computer technology that will anticipate your every move.

Its Ability Will Only Expand

The capabilities of deep learning will improve opening up even more doors to functions that will make life easier. No doubt deep learning can do amazing things today, but it is still a far cry from truly thinking like a human brain. As its technology and the science behind it grows, we can expect whole new forms of learning to be presented with models that will move away from the limitations it now holds. Today, deep learning is limited to recognition, classifications, computations, and identification. Eventually, it will be able to reason like the human mind and will be capable of abstract thinking as well. When that happens the implications that will follow will be immense.

They Will Have More Autonomy

Right now, all forms of deep learning need some involvement from humans to function. In supervised learning, the machine must be fed data and tested based on parameters set by humans. Even in unsupervised learning, these machines need some input from humans to start their program and utilize inputted data.

In the future, these machines could eventually reach the point where they can create the next generation of their technology themselves, program them completely without the aid of human interference. As they continue to learn and to store the information they have acquired, accessing it when needed, it is highly possible that this reusable knowledge will help them to design and create their own machine learning and adapt it to their own needs.

There Will be a Move Away from Models

Right now, a principal part of deep learning is the different models in existence. In the future, it is expected that the models we see in use today will one day become computer programs that will be able to adapt to all sorts of situations whether pre-programmed for it or not.

More Advancements in Medical Technology

It is feasible that every medical lab in the country will be equipped with learning machines that will be able to diagnose illnesses, test lab results, learn how to perform surgery, or dispense with medication, all without human interaction. Imagine a world where a machine can analyze your DNA samples to help doctors to learn about your potential for medical risks or to determine if you have a gene for Alzheimer's or some other illness. This knowledge would allow you to take steps ahead of time in order to stop the disease from ever happening.

Biomechanics

The future may also usher in a new era of biomechanics. Already nanotechnology is able to create limbs and other body parts for patient use. Instead of having to use a donor's heart, these tiny learning machines could be able to create a heart for you in a lab. An artificially created limb can be taught to perform like the real limbs would.

Better Mobile Technology

Can you make a smartphone any smarter? With deep learning, anything is possible. Mobile technology could literally become a personal virtual assistant, scheduling appointments, performing personal tasks, and managing your life overall. If you had the freedom away from such mundane chores you will then be free to engage in a host of other activities that you haven't been able to find the time for.

Machines Performing Mundane Tasks

One day, machines will perform the basic and more mundane tasks that humans now have to do. Self-driving trucks will come to your neighborhood will pick up your garbage every week, they will clean your home, prepare your meals, and even tidy up your home.

They Will Have Bigger Capabilities

As more algorithms are introduced, deep learning can only expand. This means there will be more automation, so each machine will be able to perform more than one task. Each layer of the neural networks can introduce a whole new feature to add to their abilities. In time, you'll have an AI that's sole responsibilities are to take care of you.

The reality is that we don't know exactly what the future holds or what to expect, but if history tells us anything, we all understand that this type of technology has a very long and prosperous future. We may not ever see the day when Sarah Conner has to fight off the Terminator and we are far from the threat of world domination by machines, but we can fully and realistically expect that machines will slowly become a major part of our lives.

We may not see it in our lifetime, but the time will come when machines will change the way live, work and play in a myriad of different ways. Every day, we are learning more about this new and exciting technology and what it can do for us. One thing for sure, our future looks bright and promising with deep learning on the horizon we have many good things to look forward to.

Machine Learning

When you start talking about deep learning, the words data science, data analytics, and machine learning will come up a lot. In fact, we have already looked a little at machine learning. A lot of people will get these terms confused, and most aren't sure which one is which. In this chapter, we will look more at the differences between all of these things so that you have a clear understanding of what they all are.

Machine learning is the practice of using algorithms to learn from data and forecast possible trends. The traditional software is combined with predictive and statistical analysis to help find the patterns and get the hidden information that was based upon the perceived data. Facebook is a great example of machine learning implementation. Their machine learning algorithms collect information for each user. Based on a person's previous behavior, their algorithm will predict the interests of the person and recommend notifications and articles in their news feed.

Since data science is a broad term that covers several disciplines, machine learning works as a part of data science. There are various techniques used in machine learning such as supervised clustering and regression. But, the data that is used in data science may not have come from a machine or any type of a mechanical process. The biggest difference is that data science covers a broader spectrum and doesn't just focus on statistics and algorithms but will also look at the entire data processing system.

Data science can be viewed as an incorporation of several different parent disciplines including data engineering, software engineering, data analytics, machine learning, business analytics, predictive analytics, and more. It includes the transformation, ingestion, collection, and retrieval of large quantities of data, which is referred to as Big Data. Data science structures big data, finding the best patterns, and then advising business people to make the changes that would work best for their needs. Machine learning and data analytics are two tools of the many that data sciences use.

A data analyst is someone who is able to do basic descriptive statistics, communicate data, and visualize data. They need to have a decent understanding of statistics and a good understanding of databases. They need to be able to come up with new views and to perceive data as visualization. You could even go as far as to say that data analytics is the most basic level of data science.

Data science is a very broad term that encompasses data analytics and other several related disciplines. Data scientists are expected to predict what could happen in the future using past patterns. A data analyst has to extract the important insights from different sources of data. A data scientist will create questions and the data analyst will find the answers to them.

Machine learning, deep learning, data science, and data analytics are only a few of the fasted growing areas of employment in the world right now. Having the right combination of skills and experience could help you get a great career in this trending arena.

Understanding Neural Networks

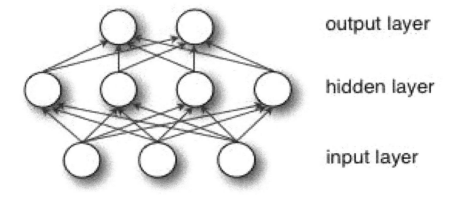

output layer

hidden layer

input layer

As we've already learned, deep learning is reliant on the use of neural networks. These networks have been designed with the human brain in mind and if you compare a diagram of the brain with one of an artificial neural network (ANN), you will see some strong similarities.

What is a Neural Network?

Basically, a neural network can be described as a system patterned to operate like the human brain. The artificial neural network (ANN) is made up of layers of interconnected neurons that can receive a series of inputs and weights. It will take the data and perform a series of mathematical computations to come up with an output or result that can be similar to that of the biological brain.

A neural network is comprised of four components:

- Neurons
- Topology – this is the connective path between the different neurons
- Weights
- Learning Algorithm

While the function of an ANN is to mimic the human brain, it cannot do so exactly. One of the reasons for this is because the ANN is designed with thousands of neurons whereas the human brain consists of billions. So, there is the possibility that it will learn in a similar way to copy the human mind, but it is far from a perfect replication of the real deal.

The Structure of a Neural Network

Still, there are many functions that an ANN can perform that can be applied in a number of ways. Let's take a look at a diagram of a biological neuron found in a human brain and compare it.

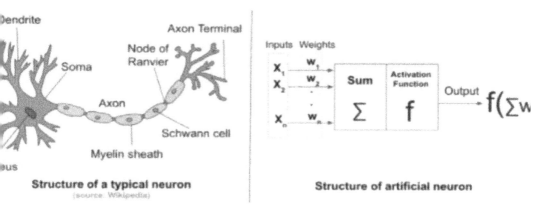

Structure of a typical neuron
(source: Wikipedia)

Structure of artificial neuron

In a biological neural network, the neurons consist of a cell nucleus that can receive input from the billions of other neurons in the network. This is done through a series of input terminals called dendrites (a group of dendrites together is referred to as a dendrite tree), which receives different types of signals from other neurons in the network. The signals can be excitatory or inhibitory in nature and are delivered via an electrochemical exchange of neurotransmitters.

How strong these input signals are will depend on several factors including the amplitude of the action coming from the previous neuron and the conductivity of the ion channels that are feeding the dendrites. The ion channels allow for the flow of electrical signals passing through the neuron's membrane or outer shell.

When signals are frequent or are of a larger magnitude, they generally have a much better conductivity in the ion channels so the signal is easier to propagate.

Depending on the type of signal received, the neuron will respond with either a message to activate or to inhibit. In other words, it will be told to turn on or turn off. Each neuron contains an electrochemical threshold, which determines whether the data received is sufficient enough to turn on the neuron or not. The result of all this activity is then sent to other neurons and the process continues.

Learning with the human brain is achieved by making tiny little adjustments to an existing configuration of neurons. Each configuration is created based on specific details before any learning can begin. How strong the connection of the neurons are, or the weights, are not random nor does the brain's topology have an effect on them.

Over time though, the strength of these connections will change based on their adjusting formation that affects both the topology and weights. Each time an adjustment is made learning happens. In the human brain, all of this happens automatically, in an instant. We watch this in babies as they learn to walk, talk, and play. Throughout our lives, we meet and get to know new people and our brains are capable of distinguishing one person from another based on numerous factors. The way they look, how they sound, their walk, etc. Every time we learn something new our brain makes these small adjustments to the neural networks and the results are stored as memories.

This function is completed in all sorts of tasks and environments. We can recognize objects, process sounds, and speech patterns. None of these skills are learned automatically but are developed slowly and repeatedly over extended periods of time. Each time the brain literally rewires itself to perfect the task it has been challenged to do.

Evidence of this was discovered through a series of experiments on animals. Their eyes were forced to remain closed for two months during their developmental stage while researchers observed the changes in the animal's visual cortex. After the two months, the animal's eyes were allowed to open but they were no longer receptive to light. In the cells and in the brain the eye had physically changed and was no longer able to function. We've now seen the same thing happening in humans. Studies have revealed that those who spend the majority of their lives in cities are found to be more sensitive to parallel lines and sharp edges, whereas those who spend the majority of their time in rural areas have eyes that are more sensitive to smooth textures.

How Learning Happens in ANNs

In artificial neural networks, the training starts with a fixed topology specifically chosen to address a certain problem. These topologies do not change with time and their weights are adjusted at random. This is done with the use of an optimization algorithm used to map the formation of input stimuli into a single cluster in order to get the desired result.

However, it is possible for ANNs to learn or to be fine-tuned based on their pre-existing representation. The process involves making adjustments to the previous weights from its original topology. This process does not happen as quickly as the human brain does but it does so at a very slow learning weight as it reacts to the newly supplied input data.

This training occurs when the weight update process starts to send data through the neural network. It measures the outcome and adjusts the weights according to the results. The weights are generally "pushed" in the direction that is most likely to improve the performance of the objective. So, an ANN designed to recognize objects will take the feedback and adjust the weights accordingly. This type of programming can be compared to a child who is learning how to recognize certain things. It is more a trial and error approach. After each failed attempt, his brain will analyze the feedback, and then take a different direction the next time.

This will be repeated until the child reaches the desired result.

An ANN works the same general way. It is first given stimuli or data that has a known response along with a learning regime that will make adjustments in an attempt to maximize accuracy.

Once the machine has learned, it can then use the experience and apply it to new stimuli, even if it is something it has never been exposed to before. The more problems the machine has to solve, the faster it will be able to learn how to tackle new ones because the connections will become much more defined.

The More Exposure – The Better

While up until recently we weren't able to explain it scientifically, we have known for centuries that the more you expose a child to the world, the faster they learn. This is true, even when the learning process is an unpleasant experience. In fact, when the learning process is painful, the feedback they receive is even more memorable. With an ANN the same could be true. When an ANN is exposed to a variety of stimuli, the fine-tuning process can ensure that it is not being overly exposed to a single thread of processing.

With each additional type of stimuli, the network can "learn" how to classify new stimuli it will receive in the future. The basic principle behind this is based on the "Black Swan Theory," in the human brain and in ANNs if they have only been exposed to one type of stimuli it is impossible for them to conceive of another type of stimuli. For centuries, people had concluded that all swans were white because there had been no recorded facts of a swan of any other color. In other words, no one had seen or been exposed to a swan that was not white.

However, later a Dutch explorer by the name of Willem de Vlamingh actually saw a group of black swans in Western Australia. Once he was exposed to this new stimuli, he had to adjust his understanding and create a new element of classifying them. If an ANN is only fed a limited stream of input, it cannot learn to classify other data that it has not been exposed to.

Its ability to draw knowledge from what it has learned is a huge stride in computer science. It allows machines to solve problems across all spectrums with a phenomenal array of applications, often with the ability to find a better solution than what would normally be achieved.

While this is a major advancement that far surpasses anything that machines have been able to do before, it is safe to conclude that this type of technology is still in its early stages. They still have a long way to go before they are capable of even coming close to being able to do what the human brain can do. Their topology is much too basic, they're learning algorithms are as of now, extremely naïve.

Still, as technology continues to advance, ANNs will continue to learn how to solve problems more effectively and face many more unknown issues in the future. Already these devices have been able to outperform human analysts in accuracy and speed in a number of areas. It is believed that one day, they will be able to multitask like humans and mimic some of the most impressive human minds in all manner of things. But that goal is now many years in the future.

Input, Hidden, and Output Layers

Now, let's look a little closer at how these neural networks work. Even someone who is not computer savvy understands that computers use processors and memory to perform complex computations. For years, man has been astounded by the speed at which these machines can calculate a result. This has been going on for decades with huge success. Think of your calculator you carry around with you all the time. Computation is a given in the world of machines. But now, because of the introduction of neural networks, a whole new approach to problem-solving has begun.

To understand this better, we must first know that a computer consists of 10^9 transistors with a switching time of 10^9 seconds. What does that mean? Transistors are tiny little switches that are triggered by electrical impulses. These are the basic building blocks used in microchips and other computer devices. Compare that to the human brain, which has 10^{11} neurons, but they have a switching time of only 10^3 seconds.

The neurons in the human brain can be compared to the human nervous system, which has three different parts. The primary part is the brain or the nerve center, which is constantly receiving input it needs to process before making decisions. Input is received and sent to the nerve center, which processes all the data, analyzes it, and sends out instructions to the rest of the body.

- Data is received from the 5 senses.
- The central nervous system analyzes it and sends out instructions on how to respond.
- The body receives instructions from the brain and

executes the appropriate action.

Now let's make another comparison with the machine to see how input differs. In a human brain:

- Synapses are at the most basic level and are completely reliant on molecules and ions for their actions.
- A neural microcircuit assigns a group of neurotransmitters to perform a particular operation
- The neural microcircuits are collected together to create subunits found in the dendrite trees
- Each neuron is capable of holding several of these dendrite subunits.
- These neural groupings work together to perform instructed operations.
- Interregional circuits create pathways, topographic maps of the cerebrum.
- **The sensory system is the point where the topographic maps interrupt certain forms of conduct.**

What is an Output Layer?

At the opposite end of the spectrum, a layer sits at the highest-level building block of the deep learning world. It is basically a container that receives the input and then adapts it with many of the non-linear functions and then passes the results on to the next layer. Layers are usually uniform in structure and can hold only one type of activation. This makes it easy to compare that layer to another part of the same network.

To make this easier to understand think of the first layer as the input layer and the last layer as the output layer.

What is a Unit?

A unit can be found in both input and output layers. It is simply the activation function where the inputs are adjusted using a nonlinear format. In most cases, a unit will have a number of income connections as well as some outgoing connections.

These neural networks calculate the weighted sum of all of the input, adjusts for bias, and then determines whether the data justifies factoring it in the decision or not. This decision is determined by checking the value that a neuron produces before deciding if an outside connection should be included.

Units are extremely complex, and a single unit can have numerous activation functions in it. The neurons contained in any given layer can receive similar functions or enactment work. The type of enactment utilized will be consistent across that specific layer.

Hidden Layers

A neural network consists of a massive number of artificial neurons (the units) that are stacked up in a series of layers. All of these neurons are interconnected through an extremely complex web. We've already discussed the input layer, which makes up the foundation of the entire network and the output layer, which performs the calculations and adapts the data to get a result. But there is most likely a number of hidden layers that lie between input and output layers.

Each neuron located in a hidden layer is also fully connected to all of the neurons in the previous layer as well as all the neurons in the following layer. It is only the input layer that does not need to modify the data received. Its sole responsibility is to receive the data from the environment. In essence, it is purely an information layer.

The hidden layer, however, must copy that data and distribute it to all the nodes it is connected to within the structure where it will perform the calculations needed to determine what to do with the information received.

The output layer performs any computations and then transmits instructions to the external environment.

Types of Neural Networks

There are several types of neural networks depending on when they were created and the extent of their complexity. The oldest of these neural networks and the simplest is the perceptron. The machine introduced in 1958, was capable of learning by means of input vectors assigned to different classes of data.

Every piece of data received has to be scaled up or down based on how important it is to the task. When a signal comes in, it is first multiplied by a predetermined weight value. So, if a neuron receives three different inputs, then that neuron will have three weights assigned to it; each can be adjusted individually.

When the computer is learning, the computer will determine the weight value based on the number of errors it made from its last test.

Next, these modified input signals are summed up to create a single value for each of the neurons. An additional computation is performed called the "bias" and is also added to the sum. After the machine learns, all the weights and biases shift so that the next result will be a little bit closer to what is the expected output demonstrating that it has actually learned from the exercise.

In the final phase, the result of the calculations is transformed into an output signal, which is then fed to the activation function and sent out to the external environment.

In perceptron, this is a very basic binary function that can produce only two possible results.

$$f(x) = \{1 \text{ if } w \cdot x + b > 0, 0 \text{ otherwise}\}$$

Based on this formula, the function will result in a 1 if the input is positive but if the input is negative, the return will be 0. Any neuron with a function like this one is a perceptron.

To train a perceptron requires the use of several preparation tests and determining the yield for each one. After each test is completed, the weights are rebalanced in a path so that it will reduce the yield error.

The Adaline neural network (ADAptive LINear Element) followed perceptron. The general rule with Adaline sometimes called the delta rule is designed to minimize the number of output errors using a gradient descent.

After a training pattern has been completed, the weights are corrected in proportion to the error percentage. The primary difference between Adaline and perceptron is the way the output applies the learning rule. In perceptron, it uses the output of the threshold function learning. However, with ADALINE, it uses the output values of -1 or +1.

The delta rule states that for a given input vector, the output vector must be compared to the correct answer. If the difference is zero, the machine did not learn. If the zero is anything else, the weights need to be adjusted as an effort to lower the difference.

The delta rule makes use of the difference between the target output values and the obtained activation. In simple terms, it compares the desired output values of the machine with the obtained activation to determine learning. It disregards perceptron's threshold activation function and instead uses a linear sum of products to calculate the activation function of the output neuron.

Throughout the training process, the strength of the connections within the network is adjusted to reduce the difference between the two values. The reason for the shift was because the threshold activation function used with perceptron could not be used in gradient descent learning needed in programs that followed. The adjustment of the values at the end of each lesson could not get progressively closer to the target with this method. However, the linear activation used with Madeline made room for error calculations to yield outputs that could be adjusted with each lesson.

How do Algorithms Work?

There is a basic principle that applies to all supervised machine learning algorithms that allow them to perform predictive modeling. The focus of each algorithm is to take the target function (f) and create a map that takes the machine from the input variable (x) to the output variable (Y). The resulting algorithm will look like this:

$$Y = f(x)$$

In this basic algorithm, the task is to make predictions about the future (Y) with examples of input variables (x).

While this formula looks on the surface to be very simple, it becomes extremely complicated when you do not know the form of function f. If this variable was known, there would be no need for the machine to learn it. Instead, it would use it in the formula to get the desired result.

There is another element that makes performing this calculation even more complicated. The algorithm must also allow for errors (e) that are completely independent of the input data (x), so the formula must be adjusted.

$$Y = f(x) + e$$

There are a number of variables that could represent e. The machine may not have enough attributes to characterize the best mapping solution. With this type of error, no matter how good the machine is at making estimates, it will never be able to reduce that error.

Algorithms like these make it possible for machines to learn and to make predictions. This is termed "predictive analytics." In general, machine learning algorithms are designed to estimate a mapping function of several output variables when they are applied to the given input variables.

With these algorithms, the machine can learn to make different assumptions based on the formula of its underlying function. If the machine does not have enough data to generate a reasonable assumption, it will not be able to learn no matter how well it performs its programmed tasks. Without algorithms carefully designed to target the task at hand, machine learning would not be what it has become today.

Neural Networks in the Future

It's hard to imagine the possibilities the future holds for neural networks. Because of how this technology is already integrating themselves into every aspect of our lives, the potential for new and innovative ideas is higher than ever. We can envision a Jetson-like world where we will have self-driving cars instead of GPS devices that need to be programmed with our intended destination. Imagine a car that has learned your personal preferences in the music you listen to, the temperature you're most comfortable with, and how to perfectly adjust your seat.

But all of that is possible now. The future holds a lot more possibilities where neural networks can be applied.

There are two different ways that this new learning technology can grow. One area is in the field of virtual intelligence. This type of program could be planned, controlled, predictable, and could eventually become the next evolutionary step in artificial intelligence.

This type of intelligence would be even closer to matching its thinking and learning styles to humans. A machine that can evolve and grow with mankind, adapt to the same environment and learn from its experiences is inevitable.

To advance to this point, however, requires technology that can actually understand and make the necessary adjustments to bridge the gap that now exists between AI and VI. As this new technology slowly engages in our world, more of our activities will be played out in virtual reality. We'll find ourselves spending more time with computers, giving the loads of data to share. We'll communicate through the use of avatars, social platforms, and games.

These virtual worlds will have to be created though, but these are places where it is safe to learn, try, and fail at our attempts to improve. They will take the place of social platforms and make it possible for us to hone our skills in business, finance, and even romance. Whatever you want to test out, there will be a virtual world to work in before you make your idea mainstream.

This will eventually become a fully automated world but not a self-aware world as many people fear. Humans will still set the parameters and put limits on the kind of things they want computers to do. Their intelligent software will be able to simplify and enhance our real life but not take charge and control it. As long as humans put limits on the computer's ability to grow, the future will remain bright for this technological advancement.

Right now, artificial intelligence is still in its infancy, the next decade could be a real eye-opener. Not very far in the future, we will begin to see these machines change the way cars and planes are designed, how they will be operated, and how they will interact with humans. We will watch our days of exploration go further and further into space. In time, we will witness the colonization of new worlds. This time literally.

The future also has many changes in store for a country's

military might. Soon, there won't be a need for "boots on the ground" when a country is at war. One soldier will be able to manage an entire fleet of drones that will fight in their place. These are already in use in some partial form now. Called unmanned aerial vehicles or UAVs these drones are capable of being operated from a remote location and responding to a myriad of instructions. In time, these UAVs will become autonomous and work without the aid of human direction.

What does this mean? Imagine a fleet of drones all headed for a single target. If one drone is destroyed by enemy fire, the remaining drones could automatically reassemble and continue on to accomplish their mission. Their ability to learn and grow will allow them to adapt to the function of the destroyed drone and incorporate his assigned task into their programming.

It is expected that with each new system introduced, machine programming will increase in its complexity and capabilities. Today, we think that artificial intelligence is one of the most fascinating forms of technology known to man. What will we think when virtual intelligence becomes available to the mainstream population? These machines will be more capable of interacting with humans and will revolutionize every aspect of our daily life.

Another area where this new technology will improve human life is in the field of disaster response. Areas unsafe for humans to enter can now be accessed by deploying machines to bring aid to people who are cut off from the rest of the world by catastrophic events. Imagine how these intelligent programs can be installed in machines that can search for life underneath the rubble of ruined buildings. How food supplies can be delivered quickly and safely. How rebuilding efforts will be much faster and how the treatment of the injured will be done quickly and efficiently.

We'll see this technology in the movie industry, music, in agriculture, and in an endless parade of other industries as time progresses. Right now, we are pretty sure of what the future holds for a neural network and all of its many applications. What we are not sure of is how quickly humankind will embrace it. No doubt, it will be the younger and more adventurous generation that will embrace it first. They will be the ones to harness its immense potential and they will be the ones who will have to set its limits.

Science has a lot to offer us in the way of advanced computer technology. The machines which will be produced tomorrow and in the years to come will open the door to a whole new world of adventure. But it will happen because people are driven by the powerful force of human desire to always find better ways to do things are stronger than the many who are powerful and in time, they will make the science fiction of the past become the reality of today.

Algebra

Linear algebra is a mathematics branch that handles vector spaces. It underpins a huge amount of data science techniques and concepts, which means that it is important to learn as much as possible.

Vectors

Vectors are objects that you can add together to make new vectors, and they can be multiplied by scalars to make new vectors as well. Vectors are points located in a finite space. While you may not view your data as a vector, they are great ways to represent numeric information.

If you are dealing with ages, heights, and weights of a large group of people, you could treat this data like three-dimensional vectors: age, weight, height. If you are teaching a class that has four exams throughout the semester, you could treat these grades as a four-dimensional vector: test1, test2, test3, test4.

One of the easiest from-scratch approaches is to show your vectors as a number list. This list of three numbers will correspond to a single vector in your three-dimensional space, and so on:

```
" height_weight_age = [70,
170,
40 ]
Grades = [95,
    80,
        75,
    62 ] "
```

A problem that comes with this approach is that you are going to want to perform some arithmetic on all of the vectors. Since Python lists don't work as vectors, and as such don't give you any tools for vector arithmetic, you will have to create these types of tools yourself. Let's see how that would work.

To start out, you will have to have two vectors. Vectors will add component-wise. All this means is that when you have two vectors, a and b, and they have the same length, they have a sum that has a first element of a[0] + b[0], and a second element of a[1] + b[1], and so on. If they don't have the same length, then they can't be added in.

If you were to add in the vectors [2, 3] and [3, 2], then you would get [2 + 3, 3 + 2] or [5, 5].

This can easily be used by zipping all of the vectors together and then making use of a comprehension to add in all of the corresponding elements.

```
" def vector_add(a, b):
Return [a_i + b_i
```

For a_i, b_i in zip (a, b)] "

In a similar manner, you can subtract your two vectors by getting rid of the corresponding elements.

" def vector_substract(a, b):
Return [a_i – b_i
For a_i, b_i in zip (a, b)] "

There may be times when you need to sum a vector list. This means that you will want to make a new vector which is the sum of the entire first elements, and the second vector should be the sum of the second elements, etc. The easiest way to do this is to take it one vector at a time.

" def vector_sum (vectors) :
Result = vectors[0]
For vector in vectors [1:] :
Result = vector_add(result, vector)
Return result "

When you really think about what we are doing, we are only reducing the vector list with vector_add. This means that we are able to rewrite this using higher-order function, such as:

" def vector_sum(vectors) :
Return reduce(vector_add, vectors) "
Or you could:

" Vector_sum = partial(reduce, vector_add) "

Next, you will find that you have to multiply your vector by a scalar. You can do this by multiplying every vector element by this number.

```
" def scarlar_multiply (c, a):
Return [c * a_i for a_i in a] "
```

This is going to give you the ability to compute the component-wise means of your same-sized vector lists.

```
" def vector_mean(vectors):
N = len(vectors)
Return scalar_multiply( 1/n, vector_sum (vectors)) "
```

One of the lesser known tools is the dot product. This product is created through the sum of two vectors and their component-wise products.

```
" def dot(a, b):
Return sum(a_i * b_i
For a_i, b_i in zip(a, b)) "
```

This product will measure how far vector a will extend in vector b's direction. One example would be if b = [1, 0], then dot (a, b) is only the first element of a. A different way to do this is by saying it is the length of the vector you would see if you were to project point a to point b.

When you use this, it becomes simple to discover the sum of the vector's squares.

```
" def sum_of_squares (a):
Return dot (a, a) "
```

And this can then be used to figure out the length or magnitude.

```
" import math
Def magnitude(a):
Return math.sqrt(sum_of_square(a)) "
```

At this point, you now have the pieces you need to figure out the space between your two vectors, as you can see in this

$$\sqrt{(a_1 - b_1)^2 + \cdots + (a_n - b_n)^2}$$
equation:

```
" def squared_distance(a, b) :
Return sum_of_squares( vector_subtract (a, b))
Def distanc(a, b) :
Return mathsqrt(squared_distance(a, b)) "
```

You can write the equivalent to get a clearer image of what we're looking at.

```
" def distanc(a, b) :
Return magnitude( vector_substract (a, b)) "
```

This is a pretty good amount of information to help you get started with vectors. It's important that you take the time to study them even further if you are still unsure of how it works.

Note: When it comes to vector lists, it works well for exposition, but it doesn't do much for performance. The NumPy library should be where you turn for production code. This library has the high-performance array classes that you will need, and it has arithmetic operations.

Matrices

These are two-dimensional number sets. The matrices will be represented as lists of lists. Each of the inner lists will have the same size and will represent a matrices row. If K is a matrix, then K[c] [d] would be the d column and c row elements. Mathematical convention dictates that matrices are represented by capital letters. You can see this here:

" K = [[1, 2, 3],
[4, 5, 6]]
L = [[1, 2],
    [3, 4],
    [5, 6]] "

Note: When it comes to mathematics, the first row of a matrix would be labeled "row 1" and the first column would be named "column 1." Since this is using a Python list, which gets indexed at zero, our first matrix row will be labeled "row 0" and the first column will be labeled "column 0."

Since we are using list-of-lists representation, our matrix K will have "len(K) rows and len(K[0]) columns." You can look at the shape.

" def shap(K) ;
Num_rows = len(K)
Num_cols = len(K[0]) if K else 0
Return num_rows, num_cols "

When you have a matrix with d columns and c rows, it is called c X d matrix. You are able to view the rows of a c X d matrix as length c's vector, and every column is the vector length d.

" def get_row(K, c) :
Return K[c]
Def get_column (K, d):
Return [K_c [d]
For K_c in K] "

You will also want to make a matrix based on the shape and a function to create the elements. This can be done through a nested list of comprehension.

" def make_matrix(num_rows, num_cols, entry_fn) :
Return [[entry_fn (c, d)
For d in range (num_cols)]
For c in range (num_rows)] "

By using this function, you can create a five-by-five identity matrix that has a 1s on the diagonal and elsewhere would be a 0s.

" def is_diagonal (c, d) :
Return 1 if c == d else 0
Identity_matrix = make_matrix (5, 5, is_diagonal)

These matrices end up being important for many different reasons.

A matrix can be used to represent a set of data that consists of several vectors by simply looking at each of the vectors as a row for your matrix. An example would be if you have the ages, heights, and weights for 1,000 people, you are able to place them in a 1,000 X 3 matrix.

" data = [[70, 170, 40],
  [65, 120, 26],
  [77, 250, 19],
  # ...
] "

You are also able to use c X d matrix to show a linear function that will map your c-dimensional vectors to your d-dimensional vectors. There are a lot of concepts and techniques that will involve these types of functions.

The third thing you can do with matrices is to use them to represent binary relationships. One representation of an edge of a network is to show them as a collection pair (c, d). But another way you could do this is to make matrix K like K[c] [d] so one of the nodes c and d are connected, and if not, they are zero.

In the former representation you would have:

" relationships = [(0, 1), (0, 2), (1, 2), (1, 3), (2, 3), (3, 4), (4, 5), (5, 6), (5, 7), (6, 8), (7, 8), (8, 9)] "

This could also be shown as:

" relationships = [[0, 1, 1, 0, 0, 0, 0, 0, 0, 0],
    [1, 0, 1, 1, 0, 0, 0, 0, 0, 0],
    [1, 1, 0, 1, 0, 0, 0, 0, 0, 0],
    [0, 1, 1, 0, 1, 0, 0, 0, 0, 0],
    [0, 0, 0, 0, 1, 0, 1, 1, 0, 0],
    [0, 0, 0, 0, 0, 1, 0, 0, 1, 0],

      [0, 0, 0, 0, 0, 1, 0, 0, 1, 0],
      [0, 0, 0, 0, 0, 0, 1, 1, 0, 1],
      [0, 0, 0, 0, 0, 0, 0, 0, 1, 0]] "

If you don't have many connections, then this wouldn't be a very efficient representation because you will more than likely have a lot of stored zeros. However, when you use a matrix representation, it will be a lot faster to check if the nodes connect. When doing this, you will only need to do this for a single matrix lookup instead of inspecting each edge.

" relationships [0] [2] == 1
relationships [0] [8] == 1 "

If you are looking to figure a nodes connection, you are going to have to look at each row or column that corresponds with the node.

" friends_of_five = [c
For c, is_friend in enumerate(relationships[5])
If is_friend] "

In the past, you may have added a connections list to all of the node objects to speed up the process, but when it comes to evolving a large graph that would end up being a bit too expensive, it would be hard to maintain.

Discrete Versus Continuous

We are going to be looking at discrete variables. Discrete variables are variables that come from a limited set. They can also include numbers with decimals depending on your variable set, but this rule has to be established. For example, if you have the number 3.578 representing the number of medical procedures that a person has had in their life, that's not possible. Even if this was just the average, it is still misleading.

You can't come out with the odds that a person has had 3.578 medical procedures in their life. They would have either had three or four. If you were looking at procedures, you would see numbers like this:

- Numbers of procedures
  - 1
  - 2
  - 3
- Odds of having that number of procedures in a year
  - 25%
  - 25%
  - 50%

When you look at continuous variables, they can't be visualized in a table. Instead, these variables have to be given in a formula as there is an infinite number of variables. An example of an input variable could be 2, 2.9, 2.99, 2.999, 2.9999 ... n.

Examples of these variables could be age, weight, and so one.

A person isn't just 32. They are typically 32 years old, 195 days, 2 hours, 1 second, 4 milliseconds. Technically, these variables could represent any single moment in time, and every interval contains infinite intervals.

Poisson Distribution

$$p(x; \lambda) = \frac{e^{-\lambda} \lambda^{x}}{x!} \quad \text{for } x = 0, 1, 2, \cdots$$

A Poisson distribution equation is used to figure out how many events could happen during a continuous interval of time. One example would be the number of phone calls that could happen during a certain time or the number of people that could end up in a queue.

This is actually a fairly simple equation to remember. The symbol is known as a lambda. This is what represents the average amount of events that happen during a certain interval of time.

An example of this distribution equation is to figure out the loss in manufacturing sheets of metal with a machine that has X flaws that happen per yard. Let's say that the error rate is two errors per yard of metal. Now, figure out what the odds are that the two errors would occur in a single yard.

Binomial Distribution

This is one of the most common and the first taught distribution in a basic statistics class. Let's say our experiment is flipping a coin. Specifically, the coin is flipped only three times. What are the odds that the coin will land on heads?

Using combinatorics, we know that there are 2^3 or eight different results combinations. By graphing the odds of getting 3 heads, 2 heads, 1 heads, and 0 heads. This is your binomial distribution. On a graph, this will look just like a normal distribution. This is because binomial and normal distributions are very similar. The difference is that one is discrete and the other is continuous.

Probability Density Function

If you have ever taken a basic statistics class, you know this function better than you think. Do you remember standard deviations? How about when you calculated the odds between the standard and average deviation? Did you realize that you were using a calculus concept known as integrals? Now, think about the space under the curve.

With this, we can assume that the space under the curve could be from negative infinity to positive infinity, or it could be a number set like the sides of a die.

But the value under the curve is one, so you would be calculating the space fewer than two points in the curve. If we go back to the sheet metal example, trying to find the odds that the two errors occur is a bit of a trick question. These are discrete variables and not continuous.

A continuous value would be zero percent.

Since the value is discrete, the integer will be whole. There wouldn't be any values between one and two, or between two and three. Instead, you would have 27% for two. If you wanted to know a value between two and three, what would the answer be?

PDF and the cumulative distribution function are able to take on continuous and discrete forms. Either way, you want to figure out how dense the odds are that fall under a range of points or a discrete point.

Cumulative Distribution Function

This function is the integral of the PDF. Both of these functions are used to provide random variables. To find the odds that a random variable is lower than a specific value, you would use the cumulative distribution function.

The graph shows the cumulative probability. If you were looking at discrete variables, like the numbers on a die, you would receive a staircase looking graph. Every step up would have 1/6 of the value and the previous numbers.

Once you reach the sixth step, you would have 100%. This means that each one of the discrete variables has a 1/6 chance of landing face up, and once it gets to the end, the total is 100%.

ROC Curve Analysis

Data science and statistics both need the ROC analysis curve. It shows the performance of a model or test by looking at the total sensitivity versus its fall-out rate.

This plays a crucial role when it comes to figuring out a model's viability. However, like a lot of technological leaps, this was created because of war. During WWII, they used it to detect enemy aircraft. After that, it moved into several other fields. It has been used to detect the similarities of bird songs, the accuracy of tests, the response of neurons, and more.

When a machine learning model is run, you will receive inaccurate predictions. Some of the inaccuracy is due to the fact that it needed to be labeled, say true, but was labeled false. And others need to be false and not true.

What are the odds that the prediction is going to be correct? Since statistics and predictions are just supported guesses, it becomes very important that you are right. With a ROC curve, you are able to see how right the predictions are and using the two parables, figure out where to place the threshold.

The threshold is where you choose if the binary classification is false or true, negative or positive. It will also make what your Y and X variables are. As your parables reach each the other, your curve will end up losing the space beneath it. This shows you that the model is less accurate no matter where your threshold is placed. When it comes to modeling most algorithms, the ROC curve is the first test performed. It will detect problems very early by letting you know if your model is accurate.

Bayes Theorem

This is one of the more popular ones that most computer-minded people need to understand. You can find it being discussed in lots of books. The best thing about the Bayes theorem is that it simplifies complex concepts. It provides a lot of information about statistics on just a few variables.

It works well with conditional probability, which means that if this happens, it will play a role in the resulting action. It will allow you to predict the odds of your hypothesis when you give it certain points of data.

You can use Bayes to look at the odds of somebody having cancer, based upon age, or if spam emails are based on the wording of the message.

The theorem helps lower your uncertainty. This was used in

WWII to figure out the locations of U-boats and predict how the Enigma machine was created to translate codes in German.

K-Nearest Neighbor Algorithm

This is one of the easiest algorithms to learn and use, so much so that Wikipedia refers to it as the "lazy algorithm."

The concept of the algorithm is fewer statistics based and more reasonable deduction. Basically, it tries to identify the groups that are closest to each other. When k-NN is used on a two-dimensional model, it will rely on Euclidian distance.

This only happens if you are working with a one norm distance as it relates to square streets, and those cars can travel in a single direction at a time. The point I'm making is that the models and objects in this rely on two dimensions, just like the classic xy graph.

k-NN tries to identify groups that are situated around a certain number of points. K is the specified number of points. There are certain ways to figure out how big your k needs to be because it is an inputted variable that the data science system or user has to pick.

This model is perfect for feature clustering, basic market segmentation, and finding groups that are among specific data points. The majority of programming languages will let you implement in a couple of code lines.

Bagging or Bootstrap Aggregating

Bagging will involve making several models of one algorithm like a decision tree. Each one of them will be trained on the different bootstrap sample. Since this bootstrapping will involve sampling with replacement, some of your data won't be used in all of the trees.

The decisions trees that are made are created with different samples, which will help to solve the problem of sample size overfitting. Decision trees that are created in this way will help lower the total error since the variance will continue to lower with every tree that is added, without increasing the bias.

A random forest is a bag of decision trees that use subspace sampling. There is only one selection of the trees features that are considered at the split of each node, which removes the correlation of the trees in your forest.

These random forests also have their own built-in validation tool. Since there is only a percentage of this data that gets used for every model, the error of the performance can be figured out using only 37% of the sample that was left by the models.

This was only a basic rundown of some statistical properties that are helpful in data science. While some data science teams will only run algorithms in R and Python libraries, it's still important to understand these small areas of data science. They will make easier abstraction and manipulation easier.

AI and Creativity

Creativity is defined as an association between two concepts that would typically not go together, but that complement each other. This could be considered something quite simple for an artificially intelligent computer. At this time, a real, living author is still needed to produce creative works of fiction, but in the future, this may not be the case.

Can Machines ever be Associated with True Creativity?

We, as humans, are familiar with machines as tools without cognitive functions comparable to that of humans. This makes it hard for people to associate them with true creativity. An issue with approaches from history was the dependence of these machines on programmed knowledge and tasks, which places an inherent limitation on what a machine is capable of. Humans have creativity because they have an awareness of physical objects and their properties. Based upon this awareness, they are able to break conventions in important ways that make sense to others.

In a similar way, technological systems need to know how to gather knowledge for the activity of creating. Research in the field of artificial intelligence is trying to get past the problem of programmed features using methods that allow these computer systems to learn, constantly, new models of information, thus modifying themselves accordingly and eventually showing true creativity.

Creativity- The Hardest Human Ability to Program:

Experts claim that robots are not likely to become creative in the near future. To automate a function, explicit and thorough instruction is required about how to accomplish creative goals. It's true that algorithms can be designed to create infinite paintings. However, it's hard, if not impossible, to teach said algorithm how to tell what makes one painting remarkable and another worthless. Another issue here is that it's hard to make automated the process of combining ideas across many sources that makes up the foundation of creativity in humans. You may, for example, find inspiration to create a poem from reading a scientific textbook. A computer cannot yet do something like this.

Are Creative Jobs being Threatened by Artificial Intelligence?

It has been said that nearly all jobs, even of a creative nature, are at a risk of becoming automated in the future. Everyone knows that our world becomes increasingly automated, leading to the loss of work for people on increasingly large scales. With a machine that has been programmed in the right way, a laborer without special skill can easily be replaced. The more research that happens, the easier it gets to pinpoint it down to algorithms and computerized formulas. With programmed artificial creative functions, machines *can* create literature and even music, just like us humans.

Artificial Intelligence and Visual Art:

Robots can also create breathtaking visual art. One remarkable program, called AARON can mix paints, wash its paintbrushes, and make masterpieces in a short period of time. Not only this, but software has added to the art of photography, letting photographs create work that would have been incomprehensible just 20 years ago. This includes a method of superimposing various pictures together to make a final image that appears to have been drawn by hand with a pencil.

What will this Mean for People with Creative Careers?

Don't worry, musicians will probably not be losing their devoted fans to robots in the near future. Later on, however, software may become an integral part of the process of creativity, aiding composures in making new songs or predicting what songs will become the most popular. We may end up being surrounded by computer-created, original songs in shopping malls, then be compelled to purchase music that was created by a computer online.

Given the fact that computers are already making news stories that are indistinguishable from a human's writing, creating masterpieces on a visual scale, and writing music enjoyed by humans, the possibilities are limitless. Since these computers are programmed and must learn about creativity through creative humans, this can increase indefinitely.

How Important is Human Emotion in Creativity?

Advances are continually made in cloud robotics. "Cloud robotics" refers to machines which are connected, in the cloud, to supercomputers. This has resulted in robots who can do surgery, cars that drive themselves, and other devices that are so advanced that they almost seem intelligent. But can robots, even those given AI, truly have creative faculties? Can art be considered art if it was created by formulas of algorithms without emotions or human thought? Sure, given how personal art is, and the fact that beauty depends upon who is interpreting the artwork. Although the creative capacity is still mostly human, this could change in the future, but given how emotional and subjective this material is, humans will probably prevail in this regard.

The Connection between Scientific Advancements and Creativity:

The connection between sudden intuitive hunches leading to amazing scientific achievements has been well-documented and mentioned. Take the famous example of a graduate student hoping to find the mathematical formula for a difficult algorithm they were studying. He was living by himself and would reflect on, study, and write ideas related to the algorithm, hoping for an answer. As a method for staying sane during this, he painted pictures of an old stone staircase outside of his flat. One morning, he woke up to realize that he could prove his answer with something called "step function". This proof has now been accepted and cited hundreds of times.

The Best Scientists are usually also Artists:

Einstein, one of the greatest minds in history, wrote about the best scientists usually also being artists. He believed strongly that the greatest of scientific achievements had to begin from intuitive understanding. Intuition can also be referred to as the sensing of a connection, a bunch, a new method for achieving something or fixing an issue. For artists, it's about discovering the right concept or idea. Although they are often considered opposite fields, art and science both rely heavily on this intuitive hunch to find out which way they should head in their work.

What Appeals to Human about a Painting Robot?

Many different reasons for creating robots that can make art exist. These range from:

Saving Time: Some, such as David Cope, believe that the time of humans can be better spent than physically composing music, writing stories, or making paintings. Why do these tasks when a robot can do them?

To Prove what Defines Human Artists: Robots can also show us that, although a robot can physically do the act of creating art, it cannot capture the subtleties, nuances, or eloquence of a real, human artist.

For Visual Appeal or Entertainment: The way people respond to watching robots create art is another part of what drives these developments in AI. It's a type of performance, and hypnotizing in a sense, for viewers to watch the robot move and create.

What Caused the Prediction of Intelligent Computers in the 1960s?

It was predicted in 1968, by Marvin Minsky, that intelligent computers would exist within one generation. In fact, he believed that they would exist, comparable to the computer HAL from the famous Stanley Kubrick film, 2001: A Space Odyssey. What caused him, along with other proponents of AI, to believe that robots could reason in such a human way?

A Fallacy in the Development of Machines:

People have always tended to believe that computers will continually get quicker, and that their kinks or clumsiness would improve exponentially, eventually leading to human-like machines. It's true that humans are progressing on this scale, when it comes to the sheer number of neurons and units of processing, leading to the prediction that we will reach levels of intelligence that are super human. But this may, actually, be quite misleading. Simply adding in more neurons, units of processing, and speed does not automatically equal a system that is more capable or intelligent.

What is necessary are new methods for understanding issues and new algorithms. When it comes to human creativity, it is not obvious that a computer that is faster will be more efficient. No matter how advanced the computer gets, it won't know what is interesting or have an ability to make connections, filtering out what is useless. The amount of neurons is irrelevant to the capacity of creativity in a machine. In this way, humans still prevail until further notice.

Is an Algorithm for Creativity Possible?

This can be possible, if it's done using variations of existing themes. For example, take the idea of a machine made to create movies. Using thousands of existing movies, it would be possible to build a machine to classify groupings or pairs of movies that already exist, to find out what works. However, the important distinction here is that it would be viewing existing patterns, not creating new ones. The machine could not discover something new that was unrelated. This is the essence of creativity; coming up with an idea that is clever and new.

The Unsettling Side of Artificial Intelligence:

Humans have a natural reaction to things when they are somewhere between dead and alive, or fictional and real. This draws a lot of interesting questions about machine and man, and comes from thinking about automation and the human organism as a functioning mechanism. The fact that people are creating automated systems to appear more and more like humans, to blur the line between real and automated, leads to anxiety about what it means to be human or real.

What Makes something Alive, or not Alive?

Plenty of literature exists around the idea of zombies, ghosts, and monsters. This idea of something bordering on alive, yet not having the full characteristics necessary to be truly human, is foundational to horror. This idea is also at the root of automation and robots. If a robot is made that looks too similar to an actual human, it turns into something unsettling. When thought goes into creating artificial intelligence that can function in the home, people wouldn't want this to appear too similar to humans. Even as advancements in artificial intelligence occur, this response remains relevant.

Examples of this Unsettling Feeling in Modern Day:

These natural, instinctive repulsions to something almost, but not quite human, are triggered in various ways. Botox is one example. When we look at a person who has had abundant injections of Botox, they have a certain unsettling appearance. Their humanness has been shifted a bit too much, leading to a natural anxiety and aversion. In the same way, humanoid androids give us this fear.

Will Robots aid Humans in Reaching their full Creative Potential?

This already occurs daily. Plentiful tools already exist for creating music and movies, much of which has benefited creativity in humans to an incredible degree. Robots and computers give us the gift of being unburdened by boring tasks like filing or looking through documents. This opens up the door for more time to be creative and improve our own artistic pursuits.

Accessing Information back Then, versus Now:

Consider the amount of time you would have had to dedicate to researching material for an article in the past. This would entail visiting a physical location (library or bookstore), walking through shelves, sifting through information. You would have to wait until a book was returned if it was already checked out by someone else.

Now we have unbelievably quick access to any information available at the click of a mouse. We can find out instantly whether the idea we just came up with is truly original, or has already been thought of. Being able to find out whether an idea has already been thought up frees up plenty of time and energy that you can devote to other pursuits. This is all thanks to artificial intelligence. So although some may feel threatened about the "takeover" of intelligent machines, we have more to thank them for than anything.

Artificial Neural Networks

The neurons of the human brain are connected together like semiconductors in a computer processor. Their interconnection creates a network in which each neuron interacts, according to a set of defined rules, with the neurons surrounding it.

An artificial neural network is a software simulation of the human brain. The network is composed of interconnected programs called neurons.

The data enters the system at the first neuron, which performs the programmed operations and passes its output to the next neuron on the network.

This neuron in turn performs its program, producing another intermediate output. Neurons in an artificial neural network are arranged in layers, and the final output is produced in the final layer.

The process of generating an output from the given input is governed by the programming of the individual neurons as well as the initial conditions of the data. Rules can be implemented in artificial neural networks so as to ensure that they simulate the human brain.

How ANNs Work

The process of learning in neural networks is implemented through the use of a learning algorithm. As we said, the neurons in a neural network are organized into layers. Each layer is connected to the layers on either side. The purpose of the input layer is to receive the input from the external environment which you need to learn about. The last layer in the neural network is referred to as the output layer, and this is the layer which gives the results of the learning process.

Between the input and the output layers, there exists some number of hidden layers. A fully connected neural network is one in which every hidden layer is connected to the layers on either side of it. The connection of one layer to another is represented by a number known as a weight. The value of the weight can be positive or negative. A higher weight will have a greater influence on the network.

The common design of a neural network in which inputs are fed into the input layer with outputs received at the output layer is referred to as a feedforward network. Each unit receives inputs from units located to its left, then multiplied by the values of the weight they have travelled along.

Every unit will add all connections it receives and if this sum exceeds a particular threshold value, will fire and propagate or trigger the units it is connected to on the right side of the network. In order for a neural network to learn, there must be feedback. The feedback is similar to a "right" or "wrong" response to the selected output or action.

This kind of learning occurs in children when they are told whether what they are doing is right or wrong -- or even in adults! Think about it. If you became aware after the fact that you failed in some task, you probably planned how you might improve your "output" the next time approach the same task. It is the same with artificial neural networks.

The purpose of feedback is to allow comparison between actual output and the desired output. The difference between the two is assessed, and a determination is made on how the process might be improved in order to generate a better output next time. The feedback process in neural networks is known as backpropagation.

The final output of the network is compared to the desired output, and the difference is used to prescribe modifications to the weights of the connections between the network units. This is done for output unit, to each of the hidden units, and lastly to the input unit. In other words, for each unit in reverse order from output to input. The process causes the neural network to learn, reducing the difference between actual and target outputs this difference is normally referred to as the network error.

The goal, obviously, is to reach a network error of zero, which requires that the network output be equal to the target output, but this may be difficult. After using several examples to train the artificial neural network, your network output will be sufficient to allow you to give the network a brand-new set of input data and observe the outputs it gives.

Consider an example in which you train your artificial neural network to differentiate between chairs and tables. You input the pictures of chairs and tables into the network and it classifies them as either chairs or tables. You train the network with 25 chairs and 25 tables. Now let say we give the network a new untested picture with perhaps a slightly different in design than earlier pictures, a designer coach chair, it should be able to classify it as either a chair or a table based on its previous experience.

However, this is not an indication that a neural network is capable of just looking at something and responding instantly to give the result. Its behavior is not like that of a person. In the example given previously, the network will not be looking at pieces of furniture. The network receives input in a binary form, meaning that each input unit is switched either on or off.

Suppose you have 5 input units, information about five

different characteristics of chairs in the form of binary answers (yes/no).

The questions can be:

Does the item pictured have a top?

A backrest?

Can you sit on it comfortably for a longer period?

Can you place a lot of stuff on it?

For a typical table, the answers will be yes, no, no, yes. In binary terms, these answers can be expressed as 1001. For a typical chair, the answers would be no, yes, yes, no. In binary terms, these can be expressed as 0110. During the learning process, the artificial neural network will learn the characteristics which represent a chair those which represent a table, and those common to both.

With this knowledge, the neural network will be able to different between chairs and tables.

Some Interesting Examples

Neural networks are widely applied today, in pattern recognition and in making decisions about patterns, for instance.

In airplanes neural networks can be used as a basic autopilot, where the input units read signals from cockpit instruments and output units which modify the controls of the plane so as to ensure that it is safe.

Artificial neural networks can be used in factories for quality control. A good example is found in the chemical process of detergent manufacturing. Detergent quality can be measured through many parameters: thickness, acidity, etc, which can be presented to a neural network as inputs. The trained network, in turn, can make a quality determination and reject or accept the batch.

Artificial neural networks are also well suited to the task of bank security. Let's say your bank runs several thousands of credit card transactions via your computer, one minute after another.

Your aim is to come up with a mechanism which will help you identify any fraudulent transactions. A neural network can help you achieve this. Possible inputs include: Is the owner of the credit card present? Has the user used a valid PIN number? Has the card transacted over 10 transactions in the last 10 minutes? Is the card being used in the country in which it was registered? Given enough such information it will be possible for the neural network to flag any fraudulent transactions, so that a human operator can be prompted to investigate further.

A bank can also use neural networks to approve or reject loan applications based upon an applicant's past credit history, employment record, and current earnings. Nearly every task involves pattern recognition and using those patterns to make decisions. Neural networks can predict everything from weather to stock market performance.

Doctors can use neural networks to assist with diagnosis. Neural networks can be used to operate radar scanning systems which can automatically identify enemy ships or aircrafts.

They can be used for voice recognition or for identification of spam emails. Neural networks have been widely used in language translation. An example of this is the automatic translation of Google, which uses neural networks to translate words from one language to another.

Neural networks have also been successfully employed to assist with police work. Applications include face recognition for use in the identification of persons or objects from captured CCTV footages.

They can also detect and recognize cars, which is helpful to police looking for stolen vehicles.

Advantages of Artificial Neural Networks

The following are the advantages of using artificial neural networks:

- Neural networks require a less formal statistical training.
- Neural networks can help in detecting complex, nonlinear relationships between dependent and independent variables. This is why neural networks are used in solving problems which do not have an algorithmic solution or where it becomes hard to find a solution.
- With neural networks, one can detect all possible interactions between predictor variables.

Disadvantages of Artificial Neural Networks

The following are the disadvantages associated with neural networks:

- Neural networks are prone to overfitting, which may lead to poor or inaccurate predictions.
- Neural networks may be slower to train and several epochs may be needed for the learning process.
- The "black box" nature of neural networks. One is only expected to provide the initial architecture of the neural network and provide some inputs to the

networks, which are just random numbers. When using backpropagation, one does not actually know what he or she is doing. The learning process of the neural network will progress on its own.

When to use Artificial Neural Networks

Neural networks should be used in problems which involve finding trends in data. They are good at solving many of the same kind problems that humans solve daily.

Examples of such problems include generalization, image recognition, and others. They help in solving problems which are difficult or impossible to solve through the use of traditional, formal analysis.

Fuzzy logic has been integrated into neural networks. This is a type of logic which recognizes more than just the simple true and false values, providing a better simulation of the real world. Suppose you have the statement, "It is sunny today."

This statement can be 100% true if there are no clouds during the day, 50% if it is hazy, 80% if there are only few clouds, and 0% if it is rainy.

This means that the concept considers outcomes such as somewhat, usually and sometimes. A combination of fuzzy logic and neural networks has been used in job application screening, automotive engineering, monitoring of glaucoma, crane control etc.

Neural networks are very good in processing massive amounts of data. This makes them very useful in image compression. More websites on the internet are using images, and neural networks can help to compress such images.

Conclusion

We have come to the end of this book. Machine learning is a branch of artificial intelligence which involves the design and development of systems capable of self-improvements showing an improvement in performance based upon their previous experiences.

In other words, these systems can "learn" by processes similar to human learning process.

Machine learning algorithms can be classified into two broad categories, supervised and the unsupervised. In supervised learning algorithms, the training data includes both inputs and outputs. The outputs (answers to the problems) are known as targets.

These in supervising the machine learning model as it tries to identify trends and patterns underlying your data.

In unsupervised learning algorithms, the training data includes inputs only. he targets are not provided. The answers to the inputs have to be discovered through a deep search.

There are a number of steps which must be followed during the course of machine learning. These include collecting and preparing the data and training, validating, and then applying the model.

When all these steps are completed, you will be able to use your model to make predictions. Machine learning is a new and growing field, and its emergence is a promising answer to the unimaginable quantities of data which will be generated by organizations and individuals during the upcoming years.

The predictive capacity of the various machine learning algorithms, is most attractive to businesses, who are rushing to incorporate machine learning into their day-to-day operations. Machine learning can help businesses predict future performance and make necessary adjustments in order to remain stable and even to increase profits.

Clearly, the future of machine learning is bright. Machine learning models can make the work of human beings easier. This fact alone should be enough to motivate human beings toward learning machine learning.

Thanks for Reading!

What did you think of, **Algorithms: Discover The Computer Science and Artificial Intelligence Used to Solve Everyday Human Problems, Optimize Habits, Learn Anything and Organize Your Life**

I know you could have picked any number of books to read, but you picked this book and for that I am extremely grateful.

I hope that it added at value and quality to your everyday life. If so, it would be really nice if you could share this book with your friends and family by posting to Facebook and Twitter.

If you enjoyed this book and found some benefit in reading this, I'd like to hear from you and hope that you could take some time to post a review. Your feedback and support will help this author to greatly improve his writing craft for future projects and make this book even better.

I want you, the reader, to know that your review is very important and so, if you'd like to leave a review, all you have to do is click here and away you go. I wish you all the best in your future success!

Thank you and good luck!

William Swain

Master key ideas in math, science, and computer science through problem solving.

Sign up for Free Now

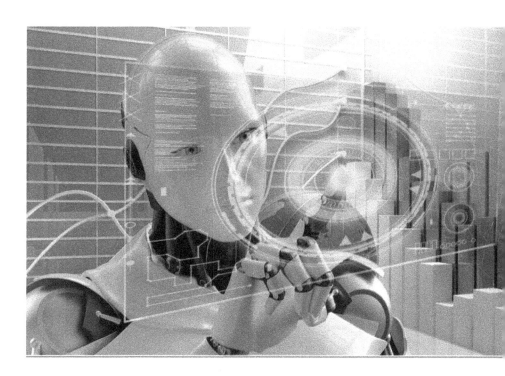

www.ingramcontent.com/pod-product-compliance
Lightning Source LLC
LaVergne TN
LVHW042335060326
832902LV00006B/193